High-Resolution Continuum Source AAS

B. Welz, H. Becker-Ross,
S. Florek, U. Heitmann

Further Titles of Interest:

J. A. C. Broekaert
**Analytical Atomic Spectrometry with Flames
and Plasmas**
2nd Edition
2005, ISBN 3-527-31282-X

E. Merian, M. Anke, M. Ihnat, M. Stoeppler
Elements and their Compounds in the Environment
Occurrence, Analysis and Biological Relevance
3 Volumes, 2nd Edition
2004, ISBN 3-527-30459-2

J. Nölte
ICP Emission Spectrometry
A Practical Guide
2003, ISBN 3-527-30672-2

B. Welz, M. Sperling
Atomic Absorption Spectrometry
3rd Edition
1999, ISBN 3-527-28571-7

High-Resolution Continuum Source AAS

The Better Way to Do Atomic Absorption Spectrometry

Bernhard Welz, Helmut Becker-Ross,
Stefan Florek, Uwe Heitmann

WILEY-VCH

WILEY-VCH Verlag GmbH & Co. KGaA

Authors

Prof. Dr. Bernhard Welz
Departamento de Química
Universidade Federal de Santa Catarina
88040-900 Florianópolis – SC
Brazil

Dr. Helmut Becker-Ross
ISAS – Institute for Analytical Sciences,
Department Berlin
ISAS – Institute for Analytical Sciences,
Department Berlin
Albert-Einstein-Strasse 9
12489 Berlin
Germany

Dr. Stefan Florek
ISAS – Institute for Analytical Sciences,
Department Berlin
Albert-Einstein-Strasse 9
12489 Berlin
Germany

Dr. Uwe Heitmann
ISAS – Institute for Analytical Sciences,
Department Berlin
Albert-Einstein-Strasse 9
12489 Berlin
Germany

All books published by Wiley-VCH are carefully produced. Nevertheless, authors, and publisher do not warrant the information contained in these books, including this book, to be free of errors. Readers are advised to keep in mind that statements, data, illustrations, procedural details or other items may inadvertently be inaccurate.

Library of Congress Card No.: applied for
British Library Cataloging-in-Publication Data:
A catalogue record for this book is available from the British Library

Bibliographic information published by
Die Deutsche Bibliothek
Die Deutsche Bibliothek lists this publication in the Deutsche Nationalbibliografie; detailed bibliographic data is available in the Internet at <http://dnb.ddb.de>.

© 2005 WILEY-VCH Verlag GmbH & Co. KGaA, Weinheim

All rights reserved (including those of translation into other languages). No part of this book may be reproduced in any form – nor transmitted or translated into machine language without written permission from the publishers. Registered names, trademarks, etc. used in this book, even when not specifically marked as such, are not to be considered unprotected by law.

Printed in the Federal Republic of Germany
Printed on acid-free paper

Printing   Druckhaus Darmstadt GmbH, Darmstadt
Bookbinding   Litges & Dopf Buchbinderei GmbH, Heppenheim

ISBN-13: 978- 3-527-30736-4
ISBN-10: 3-527-30736-2

Preface

Conventional line source atomic absorption spectrometry (LS AAS) can nowadays be considered an established technique in the positive sense of the term, i.e. it is widely used, and no dramatic improvements are expected in the foreseeable future. The state-of-the-art of conventional LS AAS is fully described in the book of Welz and Sperling [150], and its content may well be valid for another decade or two. The only progress will be in the development of new applications, but this field is today fully covered by a variety of data banks, which are easily accessible through the Internet, so that this increase in literature on applications does not justify a new edition of this book.

The only real progress in the field of AAS, in the opinion of the authors, is in the direction of high-resolution continuum source AAS (HR-CS AAS), which will undoubtedly be the future of this technique. For this reason we thought it would be much more useful to write a new book about HR-CS AAS, which might be considered a 'Volume 2' or a 'Supplement' of the above basic book on AAS. This means we expect the reader of this book to be aware of the basic concepts of AAS, which is fully covered in Reference [150], and so we have deliberately avoided repeating things in this book that have been described in the former one. For example, neither the different atomizers used in AAS, i.e. flame, graphite furnace or quartz tube atomizers, nor the atomization mechanisms or the non-spectral interferences occurring in these atomizers are discussed in this book, as they are obviously identical. In essence, only the new aspects and developments that are particular to HR-CS AAS are discussed in detail, whereas common things are repeated only where absolutely necessary.

The content of this book, regarding practical application, has essentially been produced over a time period of less than two years using prototype instruments, which are similar, but not identical, to the commercially available instrument. There have been an impressive number of people, master, doctoral and post-doctoral students, working with these prototypes, but obviously we can only give examples for the application of this new technique, not a full coverage of all the possibilities. We expect that you, the readers of this book, who hopefully will be using this exciting new technique, will be contributing to the exploration of the potential of HR-CS AAS so that the second edition of this book will contain a much more complete coverage of yet undiscovered application possibilities of this new technique.

This book is an integral part of the professorial dissertation of Uwe Heitmann. He has written several chapters and was responsible for the preparation of the figures as well as for the total arrangement and layout of this book up to the delivery of a ready-for-press manuscript. Uwe Heitmann has been concerned with the HR-CS AAS project since 1994. He was involved in most of the measurements, their evaluation and interpretation. Moreover, he carried out the setup of the prototype instruments and wrote the in-house software for data acquisition, signal processing and background correction.

Florianópolis, Berlin, December 2004

Bernhard Welz
Helmut Becker-Ross
Stefan Florek
Uwe Heitmann

Contents

1. Historical Development of Continuum Source AAS — 1

2. Theoretical Concepts — 5
- 2.1 Spectral Line Profiles — 5
  - 2.1.1 Natural Line Width — 5
  - 2.1.2 Doppler Broadening — 6
  - 2.1.3 Collision Broadening — 7
  - 2.1.4 Voigt Profiles — 8
  - 2.1.5 Instrument Profile — 11
- 2.2 Atomic Absorption with a Continuum Source — 17
  - 2.2.1 General Principle of Absorption — 17
  - 2.2.2 Instrument Effects — 18
- 2.3 Structure of Molecular Spectra — 24
  - 2.3.1 Electronic Transitions — 24
  - 2.3.2 Vibrational Spectra — 26
  - 2.3.3 Rotational Spectra — 28
  - 2.3.4 Dissociation Continua — 30

3. Instrumentation for HR-CS AAS — 31
- 3.1 Radiation Source — 31
- 3.2 Research Spectrometers with Active Wavelength Stabilization — 34
  - 3.2.1 Echelle Grating — 35
  - 3.2.2 Sequential Spectrometer — 37
  - 3.2.3 Simultaneous Spectrometer — 46
- 3.3 Detector — 50
- 3.4 The contrAA 300 from Analytik Jena AG — 53

4. Special Features of HR-CS AAS — 57
- 4.1 The Modulation Principle — 57
- 4.2 Simultaneous Double-beam Concept — 58
- 4.3 Selection of Analytical Lines — 59
- 4.4 Sensitivity and Working Range — 62
- 4.5 Signal-to-Noise Ratio, Precision and Limit of Detection — 68
- 4.6 Multi-element Atomic Absorption Spectrometry — 72
- 4.7 Absolute Analysis — 74

5. Measurement Principle in HR-CS AAS — 77
- 5.1 General Considerations — 77
- 5.2 Background Measurement and Correction — 79
  - 5.2.1 Continuous Background — 79
  - 5.2.2 Fine-structured Background — 85
  - 5.2.3 Direct Line Overlap — 89

6. The Individual Elements — 91
- 6.1 Aluminum (Al) — 94
- 6.2 Antimony (Sb) — 97
- 6.3 Arsenic (As) — 98
- 6.4 Barium (Ba) — 98
- 6.5 Beryllium (Be) — 99
- 6.6 Bismuth (Bi) — 99
- 6.7 Boron (B) — 101
- 6.8 Cadmium (Cd) — 102
- 6.9 Calcium (Ca) — 103
- 6.10 Cesium (Cs) — 103
- 6.11 Chromium (Cr) — 104
- 6.12 Cobalt (Co) — 106
- 6.13 Copper (Cu) — 108
- 6.14 Europium (Eu) — 109
- 6.15 Gallium (Ga) — 109
- 6.16 Germanium (Ge) — 110
- 6.17 Gold (Au) — 111
- 6.18 Indium (In) — 111
- 6.19 Iridium (Ir) — 112
- 6.20 Iron (Fe) — 112
- 6.21 Lanthanum (La) — 114

6.22	Lead (Pb)	115
6.23	Lithium (Li)	116
6.24	Magnesium (Mg)	117
6.25	Manganese (Mn)	117
6.26	Mercury (Hg)	120
6.27	Molybdenum (Mo)	121
6.28	Nickel (Ni)	122
6.29	Palladium (Pd)	124
6.30	Phosphorus (P)	125
6.31	Platinum (Pt)	127
6.32	Potassium (K)	128
6.33	Rhodium (Rh)	128
6.34	Rubidium (Rb)	128
6.35	Ruthenium (Ru)	129
6.36	Selenium (Se)	129
6.37	Silicon (Si)	130
6.38	Silver (Ag)	133
6.39	Sodium (Na)	133
6.40	Strontium (Sr)	135
6.41	Sulfur (S)	135
6.42	Tellurium (Te)	137
6.43	Thallium (Tl)	138
6.44	Tin (Sn)	139
6.45	Titanium (Ti)	141
6.46	Tungsten (W)	141
6.47	Vanadium (V)	142
6.48	Zinc (Zn)	144

7. Electron Excitation Spectra of Diatomic Molecules — 147

7.1	General Considerations	147
7.2	Individual Overview Spectra	153
	7.2.1 AgH	155
	7.2.2 AlCl	158
	7.2.3 AlF	160
	7.2.4 AlH	162
	7.2.5 AsO	164
	7.2.6 CN	166
	7.2.7 CS	169

7.2.8		CuH	173
7.2.9		GaCl	176
7.2.10		LaO	178
7.2.11		NH	179
7.2.12		NO	180
7.2.13		OH	185
7.2.14		PO	190
7.2.15		SH	197
7.2.16		SiO	198
7.2.17		SnO	205

8. Specific Applications — 211

- 8.1 Flame Measurements 211
  - 8.1.1 Molecular Background in Flame AAS 211
  - 8.1.2 Drinking Water Analysis 213
  - 8.1.3 Sodium and Potassium in Animal Food and Pharmaceutical Products 215
  - 8.1.4 Determination of Zinc in Iron and Steel 215
  - 8.1.5 Determination of Trace Elements in High-purity Copper 216
  - 8.1.6 Determination of Phosphorus via PO Molecular Absorption Lines 219
  - 8.1.7 Determination of Sulfur in Cast Iron 223
- 8.2 Graphite Furnace Measurements 224
  - 8.2.1 Method Development for Graphite Furnace Analysis 224
  - 8.2.2 Direct solid sample analysis 235
  - 8.2.3 Urine Analysis 237
  - 8.2.4 Analysis of Biological Materials 245
  - 8.2.5 Analysis of Seawater 251
  - 8.2.6 Analysis of Soils and Sediments 253
  - 8.2.7 Analysis of Coal and Coal Fly Ash 256
  - 8.2.8 Analysis of Crude Oil 260
  - 8.2.9 Determination of Arsenic in Aluminum 265

9. Outlook — 269

References — 273

Acknowledgment — 283

Index — 285

List of Physical Constants, Symbols and Abbreviations

Physical constant	Meaning
c	Speed of light ($2.998 \cdot 10^8$ m/s)
h	Planck's constant ($6.626 \cdot 10^{-34}$ J s)
k_B	Boltzmann's constant ($1.381 \cdot 10^{-23}$ J/K)

Symbol	Meaning [unit, as not indicated otherwise]
A	Absorbance
A_{int}	Time-integrated absorbance [s]
c_0	Characteristic concentration [μg/L]
λ	Wavelength [nm]
m_0	Characteristic mass [pg]
ν	Frequency [1/s]
τ	Lifetime [ns]

List of Abbreviations

Abbreviation	Meaning
AAS	Atomic absorption spectrometry
AC	Alternating current
ARES	Array echelle spectrograph
BC	Background correction
BCP	Background correction pixel
BOC	Background offset correction
CCD	Charge-coupled device
CRM	Certified reference material
CP	Center pixel
CS	Continuum source
DC	Direct current
DEMON	Double echelle monochromator
DSI	Dispersive slit illumination
ETV	Electro-thermal vaporization
F	Flame
FSR	Free spectral range
FWHM	Full width at half maximum
GF	Graphite furnace
HCL	Hollow cathode lamp
HFS	Hyper-fine structure
HR	High-resolution
ICP	Inductively-coupled plasma
LOD	Limit of detection
LS	Line source
MS	Mass spectrometry
OES	Optical emission spectrometry
PDA	Photodiode array
Pixel	Picture element
PMT	Photo-multiplier tube
SNR	Signal-to-noise ratio
UV	Ultra-violet
VUV	Vacuum-UV
WIA	wavelength integrated absorbance
WSA	wavelength selected absorbance

1. Historical Development of Continuum Source Atomic Absorption Spectrometry

When Bunsen and Kirchhoff [79–81] were carrying out their systematic investigation of the 'line reversal' in alkali and alkaline earth elements, i.e. the correlation between emission and absorption of radiation by atoms, in the early 1860s, they used a continuum source, i.e. 'white light', for their absorption measurements. The few researchers that used atomic absorption for their investigations in the second half of the 19th century, such as Lockyer [92], used similar equipment, as shown in Figure 1.1, for obvious reasons: Firstly, continuum light sources were the only reliable sources available at that time, and secondly, they served perfectly the purpose of detecting and measuring the 'black lines', i.e. the interruptions in the otherwise continuous spectrum, caused by atomic absorption.

In the first half of the 20th century, when atomic spectra were increasingly used not only for the qualitative identification, but also for the quantitative determination of elements, it was at least in part because of this continuum source that spectroscopists gave preference to atomic emission over atomic absorption. It is obviously much easier to detect a small radiation in front of a non-emitting, 'black' background, than a small reduction over a narrow spectral range of a strong emission. Or, if a photographic plate is used as the detector, as was common practice at that time, it is much easier to quantify a small increase in the opacity ('blackening') of the photographic layer than a small decrease in the opacity of an otherwise black plate. Hence, the radiation source was obviously the reason why atomic absorption was essentially excluded from analytical atomic spectroscopy for more than half a century.

It was only in 1952 when Alan Walsh, after having worked on the spectrochemical analysis of metals for seven years, and in molecular spectroscopy for another six years, began to wonder why molecular spectra were usually obtained in absorption and atomic

Figure 1.1: Apparatus used by Lockyer [92] for atomic absorption measurements: light source on the right; atomizer in the middle (iron tube mounted in a coal-fired furnace, while hydrogen was generated in a Kipp's apparatus to provide a reducing atmosphere); spectroscope on the left

spectra in emission. The conclusion of his musing was that there was no good reason for neglecting atomic absorption spectra [147]. Obviously, Walsh also had to consider the question of the proper radiation source for recording atomic absorption spectra, and he came to the conclusion that a resolution of approximately 2 pm would be required if a continuum source was used. This was far beyond the capabilities of the best spectrometer available in his laboratory at that time, and he concluded that 'One of the main difficulties is due to the fact that the relations between absorption and concentration depend on the resolution of the spectrograph ...' [147]. This realization led him to conclude that the measurement of atomic absorption requires line radiation sources with the sharpest possible emission lines. The task of the monochromator is then merely to separate the line used for the measurement from all the other lines emitted by the source. The high-resolution demand for atomic absorption measurements is thus provided by the line source.

Anyway, Walsh was quite fortunate because, although the hollow cathode glow had already been discovered back in 1916 by Paschen [114], and had since been used as a fine-line source for spectroscopic investigations, it was only in 1955 that the first sealed-off hollow cathode lamp was constructed [18]. Without this development and the significant amount of research that Walsh and colleagues put into the improvement of the hollow cathode lamp design, atomic absorption spectrometry (AAS) would probably not have been accepted as a routine technique to the same extent, as it has actually been. The use of modulated line radiation sources and a synchronously tuned detection system, as proposed by Walsh [146], made the AAS technique highly specific and selective, but it obviously also made it a one-element-at-a-time technique, one of the most serious limitations of conventional AAS.

However, although commercial atomic absorption spectrometers have been built exclusively according to the principle proposed by Walsh for more than four decades, research on the use of continuum radiation sources for AAS has continued throughout this period. The early publications in this field [26,30,31,38,41,43,72,103] mainly took advantage of the instability and/or low energy output of hollow cathode lamps for a number of elements, or their unavailability for other elements, particularly the rare-earth elements [31], and demonstrated in this way the superiority of the continuum source approach. Some authors, however, even questioned the validity of Walsh's approach, although the detection limits reported for those elements for which good line sources were available, were at least one order of magnitude inferior with a continuum source.

In the following years, several groups investigated wavelength modulation, using AC scanning [138], oscillating interferometers [109,145] or a combination of optical scanning and mechanical chopping [29] in order to improve the signal-to-noise ratio (SNR) and the sensitivity of continuum source AAS (CS AAS). In the latter work, Elsner and Winefordner reported analytical curves that were linear over at least three orders of magnitude, and detection limits that were close to the theoretical values [29].

A kind of turning point in this early phase of CS AAS was the work of Keliher and Wohlers [78] who for the first time used a high-resolution echelle grating spectrometer for CS AAS. The major limitation at that time was the 150 W xenon lamp used as the continuum source, which had only a relatively low energy at wavelengths below 320 nm, where most of the elements have their most sensitive lines. This work was then continued over the next 25 years by the groups of O'Haver and Harnly [42, 44–54, 90, 93, 104, 106, 107, 110, 136, 137, 157, 161], who continuously improved the system, introducing wavelength modulation [104, 161], a pulsed continuum source and a linear photodiode array detector [48, 49, 106, 107]. They also described the first, and up until now only, functional simultaneous multi-element atomic absorption spectrometer with a continuum source (SIMAAC) [44, 47, 104], and showed the applicability of this system for a variety of practical analytical problems using flame [45, 90] and graphite furnace [46, 93] atomization. The only other 'simultaneous' CS AAS instruments described in the literature [32,74] used photodiode array detectors that covered a spectral range of 2.5 nm [74] and 10 nm [32], respectively, and only elements that had absorption lines falling within this narrow spectral window could be detected simultaneously. This approach, obviously, cannot be considered a true simultaneous multi-element system.

In a review article published in 1989, Hieftje [64] provocatively predicted: 'If current trends continue, I would not be surprised to see the removal of commercial AAS instruments from the marketplace by the year 2000.' However, in the same article Hieftje also wrote: 'Clearly, for AAS to remain viable in the face of strong competition from alternative techniques will require novel instrumentation or approaches. Among the novel concepts that have been introduced are those involving continuum sources and high-resolution spectral-sorting devices ... and entirely new detection approaches.' In hindsight, this comment could be considered kind of visionary, as only one decade later, the progress made in CS AAS caused Harnly to forecast in another review article [54] that '... the future appears bright for CS AAS. Whereas, previously, CS AAS was striving for parity with LS AAS, it is now reasonable to state that it is CS AAS which is setting the standard.'

The final breakthrough in CS AAS, however, was not made by Harnly, but by the group of Becker-Ross in Berlin, who had started to work on the development of echelle spectrometers in 1980. Based on their own experience they soon discovered the weak points of the instruments used at that time [110], i.e. the low intensity of conventional xenon arc lamps in the far-UV, and the drawbacks of wavelength modulation with an oscillating quartz plate. Inspired by these ideas they started their own research in this field in 1990 [4–8, 35–37, 58, 60, 126], but with a different approach. Harnly and all the other groups essentially started from commercially available equipment and components, which they assembled and modified according to their needs. Becker-Ross and his colleagues, in contrast, first determined the requirements for CS AAS [5], and then they specified and designed the instrument according to these requirements, starting with the continuum radiation source [4, 126] followed by the spectrometer [6, 35, 36, 58] and then the detector [6, 36, 58]. All details of this concept will be discussed in detail in Chapter 3.

2. Theoretical Concepts

2.1 Spectral Line Profiles

Observed spectral line profiles are governed by a multiplicity of mechanisms, all of which cause spectral line broadening. Three mechanisms are of physical origin and act directly on atoms or molecules when generating or absorbing a photon: natural line broadening, Doppler broadening and collisional or Lorentz broadening. Another effect is of instrumental origin: broadening caused by the characteristics of the spectrometer. In this section the various broadening mechanisms and their interactions are described. The discussion will dispense with all effects of fine structure and hyperfine structure line splitting, because of their element- and line-specific character, which makes a generalized examination impossible. Moreover, except for some prominent outliers, these splitting effects are negligible in comparison to the other broadening effects.

2.1.1 Natural Line Width

Any atom being in an excited state, for instance after absorption of a photon, will undergo a relaxation process to a lower state within a finite time, even if there is no interaction with other atoms or molecules. Typical lifetimes τ for undisturbed excited states are of the order of 10^{-9} to 10^{-8} s. After this the atom re-emits the photon and relaxes to the lower state, which is the ground state in the case of resonance transitions. According to Heisenberg's uncertainty principle $\Delta E \, \Delta t = \hbar$, the finite lifetime τ causes an uncertainty of:

$$\Delta E = \frac{\hbar}{\Delta t} = \frac{h}{2\pi\tau} \qquad (2.1)$$

in the energy E of the excited state. Since the transition is associated with a photon energy of $h\nu_0 = E$, the frequency of the photon is also uncertain:

$$\Delta \nu = \frac{\Delta E}{h} = \frac{1}{2\pi\tau} \,. \qquad (2.2)$$

2. Theoretical Concepts

If the lower state is not the ground state, it will also show an energy uncertainty corresponding to its own lifetime. In this case $\Delta\nu$ is given by the sum of both contributions.

This uncertainty in frequency, which is inversely proportional to the lifetime, generates a line profile of Lorentz shape, centered at ν_0, with a width $\Delta\nu_N$. Using the relation $\Delta\lambda_N = (\lambda^2/c)\,\Delta\nu_N$ the so-called natural line width $\Delta\lambda_N$ is obtained and the corresponding wavelength-dependant intensity distribution $I_N(\lambda)$ of the area-normalized profile is given by:

$$I_N(\lambda) = \frac{1}{2\pi} \frac{\Delta\lambda_N}{(\lambda - \lambda_0)^2 + \left(\frac{\Delta\lambda_N}{2}\right)^2}, \qquad (2.3)$$

with $\lambda_0 = c/\nu_0$ and a full width at half maximum (FWHM) of:

$$\Delta\lambda_N = \frac{\lambda^2}{c} \frac{1}{2\pi\tau}. \qquad (2.4)$$

The lifetime of an electron in the excited state in the case of the resonance lines used in AAS is in the range of a few nanoseconds, resulting in $\Delta\lambda_N$ of about 0.01 pm. This is a small effect compared to the other broadening mechanisms occurring in AAS, and is therefore neglected in the context of this section.

2.1.2 Doppler Broadening

Atomic emission and absorption are always accompanied by a motion of the free atoms during each of the processes. In the case of an emission, the component of the motion in the direction of the radiation causes a frequency shift of the emitted radiation. As statistically the same number of atoms are moving in the direction of observation and in the opposite direction, the frequency shift is acting in both directions, causing a symmetric broadening of the line. In the case of an absorption process, the atoms experience a broadened frequency of the incoming radiation, and the movement of the absorbing atoms causes a further broadening of the line. Both broadening effects are due to the well-known Doppler effect. The frequency shifting effect noticed by an observer is a superposition of all contributions in the direction of the observer's view. If the atoms under consideration are in a thermodynamic equilibrium, the velocity distribution is of Maxwell type and the intensity distribution $I_D(\lambda)$ seen by the observer may be expressed by a Gaussian profile:

$$I_D(\lambda) = I(\lambda_0)\exp\left[-\left(\frac{\lambda - \lambda_0}{\frac{1}{\sqrt{4\ln 2}}\Delta\lambda_D}\right)^2\right]. \qquad (2.5)$$

$\Delta\lambda_D$, the so-called Doppler line width, is the FWHM which is given by:

$$\Delta\lambda_D = 2\sqrt{2\ln 2}\,\lambda_0\sqrt{\frac{k_B T}{c^2 m}}. \qquad (2.6)$$

If the mass m of the atom is expressed by the molar mass M given in g/mol, the width can be written as:

$$\Delta\lambda_D = 7.16 \cdot 10^{-7} \lambda_0 \sqrt{\frac{T}{M}}. \qquad (2.7)$$

Figure 2.1 shows the wavelength dependence of $\Delta\lambda_D$ for different atom masses. All values are based on a temperature of 2600 K, which is representative for an air/acetylene flame. In the most relevant region, i.e. wavelengths between 190 nm and 350 nm and masses between 14 g/mol and 200 g/mol, the variation of $\Delta\lambda_D$ is in the range 0.5 pm to 3.5 pm.

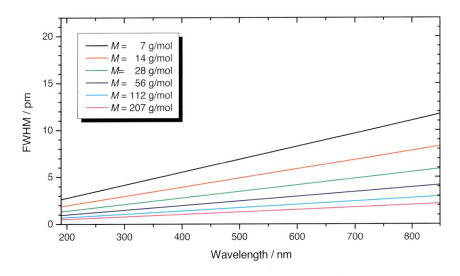

Figure 2.1: Calculated FWHM values for Doppler broadening at 2600 K and different atom masses

2.1.3 Collision Broadening

If the absorbing atoms collide with other atoms or molecules, a further broadening influence on the spectral lines is observed. A thorough discussion of the very complex collisional effects has been published by Allard and Kielkopf [1]. All of these broadening mechanisms produce a Lorentz distribution as line profile corresponding to Equation 2.3. According to Larkins [85] the collisional broadening width $\Delta\nu_C$ expressed in Hz is given by:

$$\Delta\nu_C = \frac{1}{\pi} N \sigma_C \bar{\nu}. \qquad (2.8)$$

2. Theoretical Concepts

Here, N is the perturbing atom or molecule density, σ_C is the collisional cross-section in m^2, and $\bar{v}$ is the mean relative velocity between the colliding partners. For thermal equilibrium $\bar{v}$ is given by:

$$\bar{v} = \sqrt{\frac{8 k_B T}{\pi}\left(\frac{1}{m_A} + \frac{1}{m_B}\right)}. \tag{2.9}$$

m_A and m_B are the masses of the absorbing (A) and disturbing (B) atom, respectively. For normal pressure, Equation 2.8 then transforms to:

$$\Delta \nu_C = 1.4 \cdot 10^{16}\, \sigma_C \sqrt{\frac{1}{T}\left(\frac{1}{m_A} + \frac{1}{m_B}\right)}. \tag{2.10}$$

Expressed in wavelength and by using molar masses M_A, M_B (g/mol), Equation 2.10 gives the FWHM for collisional broadening, the so-called collisional line width $\Delta \lambda_C$:

$$\Delta \lambda_C = 1.13 \cdot 10^{21}\, \lambda_0^2\, \sigma_C \sqrt{\frac{1}{T}\left(\frac{1}{M_A} + \frac{1}{M_B}\right)}. \tag{2.11}$$

Larkins determined collisional cross-sections for some elements in an air/acetylene flame and found a typical value of $\sigma_C \approx 2 \cdot 10^{-18}$ m^2. Figure 2.2 shows the wavelength dependence of $\Delta \lambda_C$ for this cross-section, a temperature of 2600 K, and different atom masses. As perturbing particle N$_2$ with $M_B = 28$ has been assumed. In the most relevant region, i.e. wavelengths between 190 nm and 350 nm and masses between 14 g/mol and 200 g/mol, the variation of $\Delta \lambda_C$ spans from 0.5 pm to 2 pm, which is comparable to the range of the Doppler broadening under the same conditions (refer to Figure 2.1).

As well as broadening, a shift of the spectral line appears, which can be towards shorter wavelengths (blue shift) or to longer wavelengths (red shift), depending on the collision partner. For the prominent case of an adiabatic impact, Corney [19] predicted the relationship between shift and broadening to be 0.36.

2.1.4 Voigt Profiles

The observable profile of a spectral line is, in general, neither a pure Lorentz nor a pure Gauss distribution but a combination of both, known as a Voigt profile. If it is assumed that Doppler and collision broadening are independent processes, the Voigt profile is the result of the convolution of the Lorentz distribution with $\Delta \lambda_C$ and the Gauss distribution with $\Delta \lambda_D$. Since the Voigt profile cannot be obtained analytically, numerical convolution procedures have to be applied. A parameter often used for profile characterization is the

2.1 Spectral Line Profiles

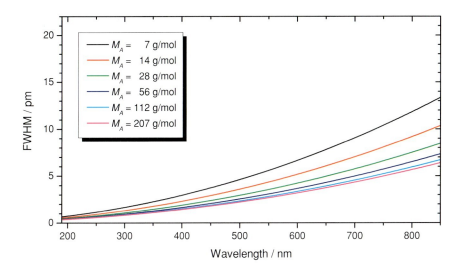

Figure 2.2: Calculated FWHM values for collisional broadening at 2600 K and normal pressure in an air/acetylene flame (perturbing particle: N_2, $M_B = 28$), curve parameter is the atom mass M_A

so-called damping constant α, which is defined as:

$$\alpha = \sqrt{\ln 2}\, \frac{\Delta\lambda_C}{\Delta\lambda_D} \,. \tag{2.12}$$

The FWHM of the Voigt profile, the so-called Voigt line width $\Delta\lambda_V$, cannot be obtained by simple addition of the Doppler and Lorentz widths, but can be approximated by an empirical formula:

$$\Delta\lambda_V \approx \frac{\Delta\lambda_C}{2} + \sqrt{\left(\frac{\Delta\lambda_C}{2}\right)^2 + \Delta\lambda_D^2} \,. \tag{2.13}$$

Figure 2.3 shows Gauss and Lorentz profiles of equal area and FWHM as well as the resulting Voigt distribution. While the Lorentz portion dominates at the line wings, the Gauss portion determines the shape in the line core.

An example of line widths in a conventional air/acetylene flame corresponding to the data in Figures 2.1 and 2.2 is shown in Figure 2.4. The widths of the Voigt profiles are calculated according to Equation 2.13. In the most relevant region, i.e. wavelengths between 190 nm and 350 nm and masses between 14 g/mol and 200 g/mol, the variation of $\Delta\lambda_V$ spans from 0.8 pm to 4.5 pm, but for longer wavelengths and the lighter elements widths of more than 10 pm could be expected.

2. Theoretical Concepts

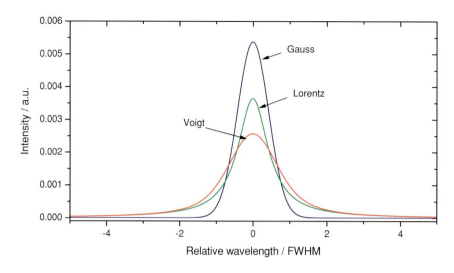

Figure 2.3: Comparison of Gauss (blue line) and Lorentz (green line) curves of equal area and same FWHM, and a Voigt (red line) profile produced by convoluting the other two curves

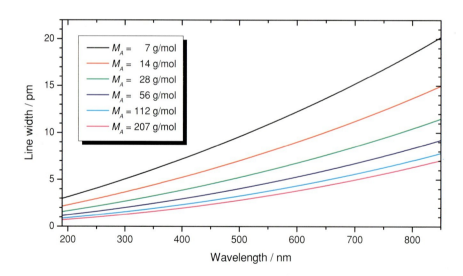

Figure 2.4: Calculated FWHM values for Voigt profiles resulting from Doppler and collisional broadening at 2600 K and normal pressure in an air / acetylene flame (perturbing particle: N_2, $M_B = 28$), curve parameter is the atom mass M_A

2.1.5 Instrument Profile

Although Voigt profiles generally describe the true line shape, every measurement of an intensity distribution by a spectrometer results in an additional influence of the instrument transmittance profile upon this shape. The geometric shape of the instrument profile is determined by the diffraction of the radiation waves, the entrance slit geometry, and the optical aberrations occurring when a monochromatic image of the entrance slit is focused onto the focal plane of the instrument. It can be described as the product of the convolution of a sinc^2 profile with a rectangular shape and a normally non-symmetric aberration distribution. The sinc^2 profile is characterized by the observed wavelength λ_0 and the relative aperture $k_{en} = f_{en}/D_{eff}$ (f_{en}: collimator focal length, D_{eff}: effective collimator diameter) of the instrument. The separation of the first sinc^2 minima from the central maximum is approximately $1.22\,\lambda_0/k_{en}$. Correspondingly, Δg_{sinc}, the FWHM of the geometric sinc^2 shape can be described as follows:

$$\Delta g_{sinc} = 1.1 \frac{\lambda_0}{k_{en}} . \tag{2.14}$$

For real entrance slit widths and negligible aberrations the geometric profile results from the convolution of the sinc^2 with a rectangle. The width of the latter, the entrance slit width s_{en}, can be chosen as $s_{opt} = 1.22\,\lambda_0/k_{en}$ to get the optimal geometric instrument profile with a FWHM of:

$$\Delta g_{instr}^{opt} = 1.37 \frac{\lambda_0}{k_{en}} . \tag{2.15}$$

For conversion of geometrical profiles to spectral profiles the geometrical width has to be multiplied by the reciprocal linear dispersion $\delta\lambda/\delta x$ of the instrument giving the width $\Delta\lambda_{instr}^{opt}$ of the optimal instrument profile:

$$\Delta\lambda_{instr}^{opt} = 1.37 \frac{\lambda_0}{k_{en}} \frac{\delta\lambda}{\delta x} . \tag{2.16}$$

Curve B in Figure 2.5 shows the corresponding profile, which can be described in good approximation by a Gauss function. Larger slit widths lead to more or less rectangular profiles, as can be observed in case a multi-pixel photo detector is used in the focal plane and the pixel width is significantly smaller than the entrance slit width. An illustration of these conditions is also given in Figure 2.5.

As an example of the validity of the above-mentioned description of the different broadening mechanisms, Figures 2.6, 2.7, and 2.8 show a comparison of calculated and measured line profiles as can be expected in a normal air/acetylene flame. In a first step the Cu doublet at 324.754 nm was measured with the high-resolution echelle spectrometer

2. Theoretical Concepts

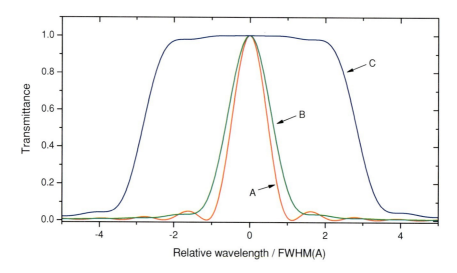

Figure 2.5: Calculated instrument profiles for three different entrance slit widths s_{en}; A: $s_{en} = 0$ (pure sinc^2 profile); B: profile A convoluted by a rectangular profile with $s_{en} = s_{opt}$; C: profile A convoluted by a rectangular profile with $s_{en} = 5\, s_{opt}$; wavelength unit: FWHM of sinc^2 profile (A)

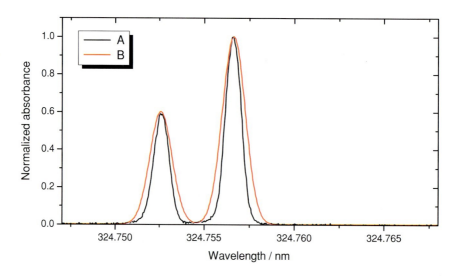

Figure 2.6: Comparison of the line profile of the Cu doublet obtained with a HCL, with a Doppler profile corresponding to 2600 K; A: lamp emission profile for 5 mA HCL current; B: Doppler profile with 1.4 pm FWHM

ELIAS II (LTB Lasertechnik Berlin GmbH, Berlin, Germany), having an instrument profile width of 0.13 pm (FWHM), which is negligible compared to the line width. The measured profiles were used to determine the line positions and intensities of the Cu doublet. For these values Gauss profiles with 1.4 pm FWHM were calculated, representing the Doppler broadening for Cu at 2600 K.

Next, the synthetic Cu profiles were convoluted by a Lorentz profile with 1.1 pm FWMH representing the collisional broadening in a 2600 K flame at normal pressure, and for an assumed collisional cross-section of $2 \cdot 10^{-18}$ m^2 (see Figure 2.2).

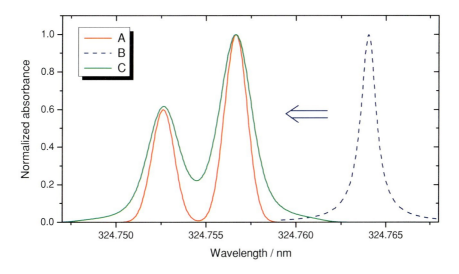

Figure 2.7: Convolution of the synthetic Doppler profile from Figure 2.6 with a Lorentz profile corresponding to 2600 K, normal pressure, and a collisional cross-section of $2 \cdot 10^{-18}$ m^2; A: Doppler profile, B: convoluting Lorentz profile with 1.1 pm FWHM, C: resulting Voigt profile

In a last step, the arising profile was convoluted by a Gauss function with 0.9 pm FWHM, representing the instrument profile of the below-discussed spectrometer SuperDEMON (refer to Chapter 3.2.2) at 324 nm. The calculated profile is faced to an absorbance profile measured in an air/acetylene flame with the SuperDEMON. The comparison shows a satisfying conformity of calculated and measured profiles.

With a numerical deconvolution procedure, corresponding to the convolution method demonstrated above, Voigt widths for some measured analytical lines were determined. The results are collected in Table 2.1 and compared with calculated values corresponding to Equations 2.7, 2.11, and 2.13.

2. Theoretical Concepts

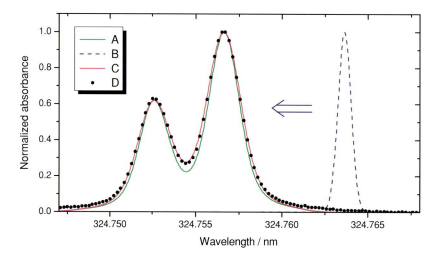

Figure 2.8: Convolution of the synthetic Voigt profile from Figure 2.7 with a Gauss profile representing the instrument function of SuperDEMON; A: Voigt profile; B: convoluting Gauss profile with 0.9 pm FWHM; C: convolution result; D: measured absorbance profile in an air / acetylene flame

Table 2.1: Comparison of measured with calculated FWHM values for Lorentz profiles

Element	Wavelength / nm	Measured $\Delta\lambda_V$ / pm	Calculated $\Delta\lambda_V$ / pm	Ratio measured / calculated
Ag	328.068	1.71	1.77	0.97
Ag	338.289	1.74	1.84	0.94
Ca	422.673	3.39	3.60	0.94
Cd	228.802	1.10	1.07	1.03
Co	240.725	1.73	1.47	1.17
Cr	357.868	2.18	2.60	0.84
Cu	324.754	2.02	2.10	0.96
Cu	327.396	1.99	2.10	0.95
Fe	248.327	1.51	1.56	0.97
Ga	287.424	2.41	1.73	1.40
Ga	294.364	2.19	1.78	1.23
Ga	294.417	2.02	1.79	1.13
Ga	403.299	4.54	2.74	1.66

Table 2.1 (continued)

Element	Wavelength / nm	Measured $\Delta\lambda_V$ / pm	Calculated $\Delta\lambda_V$ / pm	Ratio measured / calculated
Ga	417.204	3.51	2.88	1.22
In	303.936	2.15	1.55	1.39
In	325.609	2.35	1.71	1.38
Ir	208.882	0.79	0.78	1.02
Ir	237.277	1.21	0.92	1.31
Ir	250.298	1.24	1.00	1.24
K	404.414	5.78	3.43	1.69
K	404.721	5.42	3.43	1.58
K	766.491	9.69	8.75	1.11
K	769.897	9.42	8.81	1.07
Li	323.266	6.76	5.54	1.22
Li	670.785	12.04	14.37	0.84
Mg	285.213	2.54	2.68	0.95
Mn	279.482	2.41	1.83	1.31
Mo	313.259	1.48	1.73	0.86
Na	330.237	4.44	3.28	1.35
Na	330.298	4.32	3.29	1.31
Na	588.995	7.42	7.13	1.04
Na	589.592	7.20	7.15	1.01
Pb	217.001	1.19	0.80	1.49
Pb	283.306	1.80	1.16	1.55
Pd	244.791	1.15	1.19	0.96
Pd	247.642	1.19	1.21	0.98
Rh	343.489	1.63	1.91	0.85
Ru	349.895	1.73	1.97	0.88
Sb	217.581	1.47	0.97	1.51
Se	196.026	1.27	1.01	1.25
Se	203.395	1.11	1.06	1.05
Se	206.279	1.22	1.08	1.13
Se	207.480	1.15	1.09	1.06
Sn	224.605	1.15	1.02	1.13
Sn	286.332	1.67	1.41	1.18
Sr	460.733	3.13	3.08	1.02

2. Theoretical Concepts

Table 2.1 (continued)

Element	Wavelength / nm	Measured $\Delta\lambda_V$ / pm	Calculated $\Delta\lambda_V$ / pm	Ratio measured / calculated
Te	214.281	1.09	0.93	1.17
Tl	276.787	1.77	1.12	1.57
Zn	213.856	1.17	1.22	0.96

Another comparison of measured and calculated Voigt profiles is shown in Figure 2.9. The calculated line widths are displayed, as in Figure 2.4, with the molar mass M_A as parameter. The measured values are grouped by the masses of the respective absorbing atoms. The classification is indicated by different colors and is as follows: black $M_A = 7$, green $M_A = 23\ldots 40$, blue $M_A = 52\ldots 79$, cyan $M_A = 88\ldots 128$, and magenta $M_A = 192\ldots 207$. As can be expected, the significant deviations of the measured from the calculated values are directed to larger line widths, which should be caused by non-resolvable line splitting. For the majority of lines, however, the agreement is satisfactorily and demonstrates the usefulness of the presented approximations.

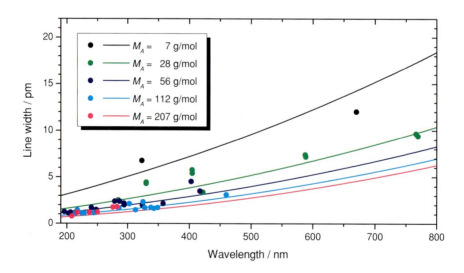

Figure 2.9: Comparison of measured (dots) and calculated (curves) Voigt widths for the absorption lines given in Table 2.1, grouped by atom masses corresponding to values used as parameters for calculated curves

2.2 Atomic Absorption with a Continuum Source

2.2.1 General Principle of Absorption

The basic description of the attenuation of radiation by passing an absorbing volume is given by Beer's law:

$$\phi = \phi_0\, e^{-kl} . \tag{2.17}$$

The transmitted radiant power will be diminished exponentially with increasing absorption coefficient k as well as with increasing length l of the absorbing layer. If the beam of a continuum radiation source is passed through an absorption cell containing a one-component mono-atomic gas with an absorption line at wavelength λ_0, the transmitted radiant power ϕ_λ becomes a function of the wavelength, as depicted in Figure 2.10.

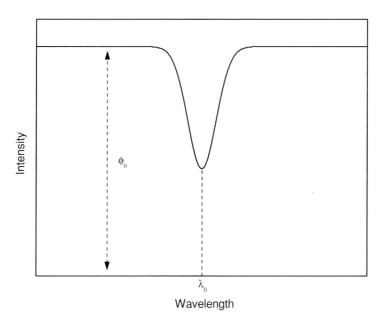

Figure 2.10: Continuum spectrum with a single absorption line

The spectrum of the transmitted radiant power shows the absorption line of the gaseous atoms centered at wavelength λ_0, and the absorption coefficient in Beer's law becomes dependent on the wavelength:

$$\phi_\lambda = \phi_0\, e^{-k(\lambda)l} . \tag{2.18}$$

2. Theoretical Concepts

The shape of $k(\lambda)$ is determined by the effects of Doppler and collisional broadening (refer to Section 2.1). For the quantity one can use the results of the classical radiation theory [105], which provides the correlation between $k(\lambda)$ and N, where N denotes the volume concentration of absorbing atoms in the ground state:

$$\int k(\lambda)\,\mathrm{d}\lambda = \frac{\pi e^2 \lambda_0^2 f N}{mc^2} \sim N\,. \tag{2.19}$$

The factor f represents the so-called oscillator strength given by quantum mechanical relationships; the symbols e and m denote the electron charge and the electron mass, respectively. In order to obtain a result directly proportional to the number of atoms, the radiant power absorbed at the absorption line has to be measured in AAS, i.e. the incident (ϕ_0) as well as the transmitted (ϕ_λ) radiant power have to be ascertained. Equation 2.18 can be modified as follows:

$$A \equiv \log\left(\frac{\phi_0}{\phi_\lambda}\right) = \log(e)\,k_\lambda\,N\,l\,. \tag{2.20}$$

The dimensionless number for absorbance A is also called optical density. Standardized line profiles may be used as function k_λ. The absorbance A is proportional to the number of atoms in a volume with concentration of absorbing atoms N, and length l.

2.2.2 Instrument Effects

In this section the influence of the spectrometer used on the result of an absorbance measurement will be discussed. Thereby the minimum detectable absorbance and the sensitivity will be investigated from the point of view of a variable width of the instrument profile. For a clear description, a virtual instrument will be assumed. This instrument might contain an entrance slit with variable width s_{en}, and a linear multi-pixel photo detector with m pixels and the single pixel width w_{pix}. The relations $w_{\mathrm{pix}} \ll s_{\mathrm{en}}$ and $m\,w_{\mathrm{pix}} > s_{\mathrm{en}}$ are assumed to hold. All pixels of the detector will be illuminated simultaneously, i. e. signals of different pixels are completely time-correlated. Furthermore, a pure rectangle is assumed for the shape of the instrument profile, hence the sinc^2 component (refer to Section 2.1.5) of the profile generated by diffraction will be neglected. A scheme of such a virtual CS AAS spectrometer is depicted in Figure 2.11.

For small absorbance values, i.e. low concentrations of absorbing atoms N or, in other words, close to the limit of detection (LOD) the demand on the measurement of ϕ_0 as well as of ϕ_λ can be reduced to the simple task: Determination of a small reduction of a large signal. Many papers on this task have been published. The crucial clarification was given by Snelleman in 1968 [138]. He developed algorithms, which clearly and straightforwardly describe the influence of the lamp property and the spectrometer characteristics on the

2.2 Atomic Absorption with a Continuum Source

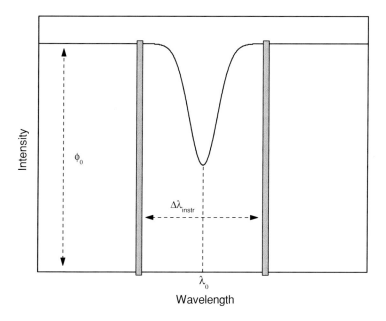

Figure 2.11: Virtual CS AAS spectrometer with rectangular instrument profile

amount of shot-noise. He pointed out, that the absorbance SNR is independent of the spectrometer bandwidth. His argumentation is adopted in the following paragraphs.

The radiant power ϕ_λ on the detector of a monochromator, with a proper imaging of a continuum source (CS), is quadratically proportional to the bandwidth, resulting in an inverse proportionality of the shot-noise dominated absorbance noise. These facts are mathematically described in the following way:

$$\phi_\lambda = L_\lambda \, \tau \, \Delta\lambda_{\text{instr}} \, \frac{s_{\text{en}} \, h_{\text{en}} \, D_{\text{eff}}}{f_{\text{en}}^2} \, . \tag{2.21}$$

L_λ denotes the radiance of the CS, τ is the transmittance of the instrument, and s_{en} and h_{en} are the width and the height of the entrance slit. D_{eff}, f_{en} and $\Delta\lambda_{\text{instr}}$ have been introduced in Section 2.1.5. By using the angular dispersion $\delta\beta/\delta\lambda$ of the dispersive optical element (which usually is the grating) s_{en} can be substituted by:

$$s_{\text{en}} = \Delta\lambda_{\text{instr}} \, f_{\text{en}} \, \frac{\delta\beta}{\delta\lambda} \, . \tag{2.22}$$

Equation 2.21 can now be written as follows:

$$\phi_\lambda = L_\lambda \, \tau \, \Delta\lambda_{\text{instr}}^2 \, \frac{h_{\text{en}} \, D_{\text{eff}}}{f_{\text{en}}} \, \frac{\delta\beta}{\delta\lambda} \, . \tag{2.23}$$

2. Theoretical Concepts

These are the well-established proportionalities for the radiant power measured with a CS. Because of the statistical character of the mean photon flux per second $n = \phi_\lambda \lambda / (hc)$, the standard deviation of n, observed during a time interval t, equals $\sqrt{n\,t}$. The minimum detectable reduction of n during observing time t, corresponding to a detectable decrease in the radiant power, is defined as:

$$\Delta n_{min} = C \sqrt{2} \sqrt{n\,t} . \qquad (2.24)$$

Here, C is the reliability factor, which is usually assumed to equal three. The factor $\sqrt{2}$ takes into account that each absorbance measurement demands two independent intensity measurements. If one examines α, the fraction of absorbed incident radiant power, the minimum detectable fraction α_{min} is given by:

$$\alpha_{min} = \frac{\Delta n_{min}}{n\,t} = 3\sqrt{2}\,\frac{1}{\sqrt{n\,t}} . \qquad (2.25)$$

According to the relationship between n and ϕ_λ, Equation 2.25 can be written as follows:

$$\alpha_{min} = 3\sqrt{2}\,\frac{1}{\sqrt{\phi_\lambda \frac{\lambda}{hc} t}} . \qquad (2.26)$$

On the other hand, the relative radiant power reduction α generated by volume absorption liable to Beer's law can be written as:

$$\alpha = \frac{\int \phi_0\,d\lambda - \int \phi_0\,e^{-k_\lambda l}\,d\lambda}{\int \phi_0\,d\lambda} . \qquad (2.27)$$

If ϕ_0 is constant over the measuring interval $\Delta\lambda_{instr}$ and if $(-k_\lambda l)$ is small compared to one, the integration over $\Delta\lambda_{instr}$ results in:

$$\alpha = \frac{\Delta\lambda_{instr} - \int (1 - k_\lambda l)\,d\lambda}{\Delta\lambda_{instr}} , \qquad (2.28)$$

and in a next step one gets:

$$\alpha = \frac{l \int k_\lambda\,d\lambda}{\Delta\lambda_{instr}} . \qquad (2.29)$$

Using equation 2.19, the following proportionality is achieved:

$$\alpha \sim \frac{l\,N}{\Delta\lambda_{instr}} . \qquad (2.30)$$

2.2 Atomic Absorption with a Continuum Source

The concentration N_{min} at the shot-noise dominated LOD can be calculated from Equations 2.23 and 2.26 by setting $\alpha = \alpha_{min}$:

$$N_{min} \sim \frac{\alpha_{min} \Delta\lambda_{instr}}{l} = 3\sqrt{2} \frac{\Delta\lambda_{instr}}{l \sqrt{\phi_\lambda \frac{\lambda}{hc} t}}$$

$$= 3\sqrt{2} \frac{1}{l \sqrt{L_\lambda \tau \frac{h_{en} D_{eff}}{f_{en}} \frac{\delta\beta}{\delta\lambda} \frac{\lambda}{hc} t}} . \qquad (2.31)$$

Equation 2.31 exemplifies that the shot-noise dominated LOD for an absorbing atom concentration can be reduced only by an enhanced radiance of the lamp, an enlarged transmittance and geometrical conductance of the spectrometer, an increased angular dispersion of the dispersing optical component, and/or an increased measuring time. All these improvements, however, produce square-root effects only.

Indeed all these conclusions have been made assuming the instrument profile to be significantly broader than the line width of the absorbing line. Therefore it should be interesting to analyze the situation for small $\Delta\lambda_{instr}$ values. The virtual spectrometer introduced above was used to model the absorbance generated by a Voigt-shaped line, which is a convolution of a Gauss and a Lorentz profile with equal FWHM. In order to describe realistic measurement conditions, a 3 % stray light level is assumed. In Figure 2.12 resulting transmittance curves for profiles with different peak absorbance values are shown.

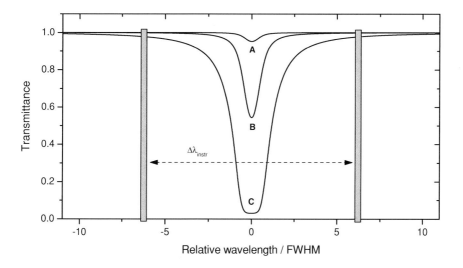

Figure 2.12: Calculated transmittance profiles for Voigt-shaped absorption lines with maximum absorbance values of (A) 0.03, (B) 0.3, and (C) 3, assuming 3 % stray light level

2. Theoretical Concepts

The results of a computer simulation of the above-mentioned virtual CS AAS measurement corresponding to Figure 2.12 are presented in Figure 2.13. The calculated curves show, for different widths of a symmetrical rectangular instrument profile, the relationship between the integrated absorbance, averaged over $\Delta\lambda_{instr}$, and the maximum absorbance. The latter is calculated for vanishing stray light and infinitesimal instrumental width. Voigt functions are used as line profiles. The width of $\Delta\lambda_{instr}$ is expressed in multiples of the FWHM of the normalized Voigt curves. As expected, for $\Delta\lambda_{instr} = 0.2 \cdot \text{FWHM}$, the averaged absorbance values and the line center values are identical up to the stray light level of 3 %, corresponding to an absorbance of 1.5. For $\Delta\lambda_{instr} = 2 \cdot \text{FWHM}$ the sensitivity in the low absorbance range is approximately a factor of two smaller than the ideal sensitivity. For values $> 2 \cdot \text{FWHM}$, the influence of the instrumental width on the shape of the curves becomes obvious, showing a hyperbolic characteristic, as predicted by Harnly and co-workers [51].

A second fundamental question arising with small instrument width is the dependence of the SNR on $\Delta\lambda_{instr}$. Calculating α numerically and using the result of Equation 2.26, which indicates the inverse proportionality of shot-noise determined α_{min} and $\Delta\lambda_{instr}$, one gets the relationship between A_{min} and the instrumental width, since A_{min} is related to α_{min} by $A_{min} = -\log(1 - \alpha_{min})$.

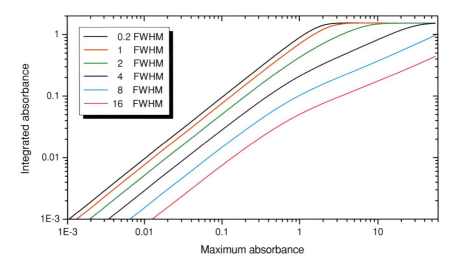

Figure 2.13: Calculated relationship between wavelength-integrated absorbance and peak (maximum) absorbance in the line center for Voigt-shaped absorption lines and a rectangular instrument profile; curve parameter: $\Delta\lambda_{instr}$, assumed stray light level: 3 %

2.2 Atomic Absorption with a Continuum Source

Figure 2.14 shows calculated signal-to-noise values of the relative absorbance. The curves were normalized to their maximum. In addition to the expected tendency that SNR values approach zero for $\Delta\lambda_{instr} \ll$ FWHM (absorption line), because no absorbance measurement should be possible without intensity, there is an interesting difference between Gauss and Lorentz profiles. The significant maximum for the Gauss-determined curve is caused by the fact that, for greater distances from the line center, the Gauss function reduces to extremely small values, so that the wings of the profile do not carry any information. In contrast, the Lorentz dominated curve looks like the right-hand part of a Lorentz shaped absorption profile.

Altogether, the results deliver a clear answer to the question, what is the optimal instrumental bandwidth for CS AAS. The bandwidth should never be chosen smaller than two times the FWHM of absorbing lines. Higher values for $\Delta\lambda_{instr}$ will not significantly reduce the shot-noise limited A_{min}. Nevertheless they should be avoided to allow the best possible background correction (BC). With respect to the results presented in Section 2.1.5, an optimized value for the instrumental resolving power $R = \lambda / \Delta\lambda$ is between 50 000 and 150 000.

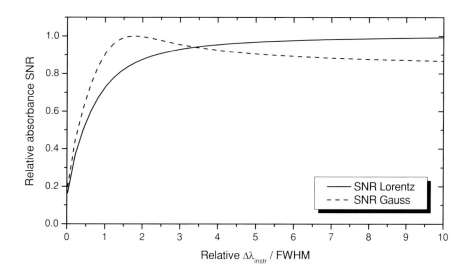

Figure 2.14: Normalized, shot-noise determined, minimum detectable absorbance for Gauss- and Lorentz-shaped absorption lines as function of $\Delta\lambda_{instr}$; rectangular instrument profile assumed

Last but not least, the influence of the detector characteristics on the absorbance signal shall be evaluated qualitatively. The whole measurement signal I is generated by photo-

2. Theoretical Concepts

electric detection of the radiant power ϕ. It can be expressed as a number of photoelectrons, if η_Q denotes the quantum efficiency of the detector:

$$I = \phi \eta_Q t .\tag{2.32}$$

The uncertainty ΔI of I is principally determined by three effects: detector read-out noise $\Delta I_{\text{det}} = $ constant, radiant power flicker-noise $\Delta I_{\text{flick}} \sim I$, and shot-noise of the signal $\Delta I_{\text{shot}} = \sqrt{I}$. The resulting noise ΔI_{res} is given by the well-known operation:

$$\Delta I_{\text{res}}^2 = \Delta I_{\text{det}}^2 + \Delta I_{\text{flick}}^2 + \Delta I_{\text{shot}}^2 .\tag{2.33}$$

For state-of-the-art solid state detectors, as for instance CCD arrays, the read-out noise is in the region of 5 to 30 electrons per read-out, i.e. for I values larger than 2000 electrons ΔI_{det} becomes negligible compared to shot-noise. With respect to flicker-noise one can take into account the time-correlated signal generation. The detector records a spectral line and its vicinity simultaneously and delivers an ideal monitoring of all flicker effects. If this advantage is exploited consistently for the calculation of absorbance, the flicker-noise becomes negligible as well. Finally, one gets $\Delta I_{\text{res}} = \Delta I_{\text{shot}}$, and $1/\sqrt{\eta_Q}$ becomes an additional factor in Equation 2.31.

2.3 Structure of Molecular Spectra

Molecular spectra consist of transitions between different molecular states, each of them having its own specific energy. Due to the complexity of a molecular system and its additional internal degrees of freedom (vibration and rotation), the number of possible states is far larger than the number found in atoms. Consequently, molecular spectra are comprised of many more lines than atomic spectra, and/or non-structured bands over relatively large wavelength ranges. This section gives a short introduction to the structure of these spectra, including main transition types, energies and wavelengths. It is the primary intention here to provide supplementary information about the topic outlined in Chapter 7.

2.3.1 Electronic Transitions

Basically there are three types of energy, which may be 'stored' in a molecule. In descending order of energy content they are electronic (E_{el}), vibrational (E_{vib}), and rotational (E_{rot}) energy. The former is based upon the Coulomb interaction between nuclei and electrons, in accordance with atomic energies. Vibrational and rotational energy are due to internal motion of the nuclei. Such contributions are small compared to electronic energy and will be treated in Sections 2.3.2 and 2.3.3.

2.3 Structure of Molecular Spectra

In a stable molecular configuration, the outer atomic orbitals overlap and create molecular orbitals which are, at least in part, occupied by electrons. The energy differences between such orbitals are comparable to atomic energies. Therefore, molecular electronic transitions, and the resulting electron excitation spectra, occur at energies or wavelengths similar to those found in atomic transitions, i.e. in the UV or visible range, often resulting in overlaid spectra of both types. A diatomic molecule scheme is shown in Figure 2.15, which demonstrates an electronic transition to a higher energy orbital induced by absorption of a photon at wavelength λ.

An electronic molecular transition is labeled by the main attributes of the initial and the final electronic state. For diatomic molecules, which play a major role in AAS, the labeling usually includes the following information, given in order of appearance: (i) the energetic order of the electronic state, (ii) the total spin of the electrons, (iii) the total angular momentum of the electrons with respect to the molecular axis, and (iv) symmetry properties of the electronic wave function, whose squared value describes the probability of finding the electrons at any location within the molecule. A typical electronic state may therefore be labeled by:

$$X/A/B/\ldots \ ^{2S+1}\Lambda_{g/u}^{+/-} \ .$$

The leading letter is either 'X' for the lowest electronic state (ground state) or 'A', 'B', ... for electronically excited states, the latter being replaced by small letters in the case of electronic states with a resulting spin $S > 0$. S denotes the total spin $(0, \frac{1}{2}, 1, \ldots)$, Λ is the total angular momentum expressed by capital Greek letters Σ, Π, Δ, ... for $\Lambda = 1$, 2, 3, ... The optional subscript 'g' or 'u', which only occurs for homo-nuclear diatomic molecules, describes the behavior of the electronic wave function in the case of inversion at the molecular center (with or without changing its sign), and finally the superscript '+' or '−' denotes the behavior of the wave function in the case of reflection at a plane including the molecular axis (with or without changing its sign).

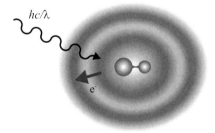

Figure 2.15: Scheme of a diatomic molecule undergoing an electronic transition by absorption of a photon with energy hc/λ

2. Theoretical Concepts

Figure 2.16 gives an example of the nomenclature for the absorbance spectrum of PO, observed in a flame (see Section 7.2.14). In the analytically important wavelength range between 200 nm and 350 nm, three electronic transitions to the excited states D, A, and B can be observed. Also visible is the striking substructure of these transitions, which will be discussed in the next sections.

As typical energies of electronic states are of the order of a few eV and the thermal energy of the absorber is in the range of 0.2 eV, nearly all atoms remain in their ground electronic state. Hence, almost every observable electronic transition starts from the X-state.

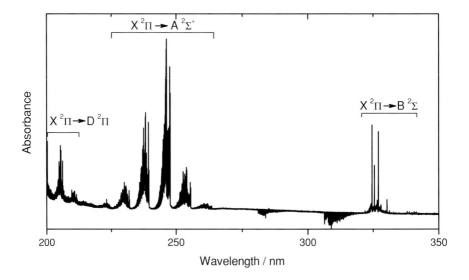

Figure 2.16: Nomenclature of the electronic transitions demonstrated for the absorbance spectrum of the diatomic molecule PO observed in a flame

2.3.2 Vibrational Spectra

Any molecule has internal energy due to vibrations of the nuclei with respect to their equilibrium configuration. The amount of vibrational energy E_{vib} is typically one to two orders of magnitude less than the electronic energy E_{el} described above. Hence, pure vibrational transitions are observed in the IR region (classical IR spectroscopy). The total energy E_{tot} of the molecule is approximately given by the sum of both energies:

$$E_{tot} = E_{el} + E_{vib} \ . \tag{2.34}$$

2.3 Structure of Molecular Spectra

For diatomic molecules the only possible vibration is a stretching of the intra-nuclear axis, as depicted in Figure 2.17. For a purely harmonic motion of the nuclei, E_{vib} can be described as:

$$E_{\text{vib}} = h\nu \left(v + \frac{1}{2} \right) \quad \text{with} \quad v = 0, 1, 2, \ldots . \tag{2.35}$$

Equation 2.35 shows that the allowed vibrational energies are equally spaced and separated by $h\nu$. The amount of the latter is specific to the nuclei and the strength of their coupling. The number v is called the vibrational quantum number. It denotes the amount of vibrational excitation energy in the respective molecular state.

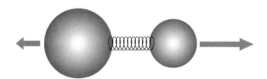

Figure 2.17: Stretching vibration of a diatomic molecule

Within an electronic transition the initial and the final state may possess additional vibrational energy, according to equation 2.34. The resulting absorbance spectrum then exhibits a substructure caused by allowed vibrational transitions. These are given by:

$$\Delta v = v(A) - v(X) = 0, 1, 2, \ldots \tag{2.36}$$

for an electronic transition from X to A. A single vibrational transition $v(X) \rightarrow v(A)$ is called a band, several bands with a constant Δv value are referred to as a sequence. As a general rule, the probability for vibrational transitions decreases for increasing Δv. Observable transitions are therefore restricted to low values of Δv. As the thermal energy of the absorber is roughly comparable to a vibrational energy given by $v = 0$, only those vibrational states with small values of v are populated, leading to a further restriction of the observable vibrational transitions.

A spectrum of an electronic transition with underlying vibrational structure is shown in Figure 2.18. Altogether six sequences, from $\Delta v = +3$ to $\Delta v = -2$, can be observed within this region of the PO absorbance spectrum. In accordance with the equal spacing of the vibrational energies, all observable transitions with $\Delta v =$ constant are strongly overlapping, and consecutive sequences are properly aligned. The substructure clearly visible within the band systems is addressed in the next section.

2. Theoretical Concepts

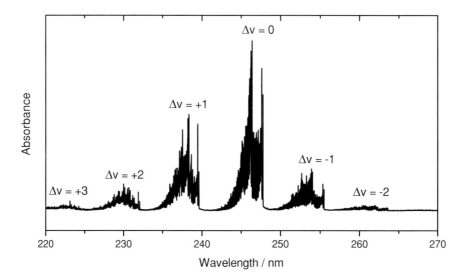

Figure 2.18: Example of vibrational transitions with equidistant energy spacing within the same electronic transition (PO molecule, refer to Figure 2.16)

For polyatomic molecules the number of possible vibrations increases. The thermal energy of the absorber then populates many more vibrational states, leading to an increased number of observable bands within an electronic transition.

2.3.3 Rotational Spectra

Another type of internal energy is given by molecular rotation. From a classical point of view, any non-linear molecule may rotate around three mutually perpendicular axes, all of them intersecting the molecular center of mass. Rotational energies E_{rot} are usually two to three orders of magnitude lower than vibrational energies E_{vib}, and the corresponding pure rotational spectra are observable in the far-IR or microwave region. Altogether the following relation holds for the three molecular types of energy:

$$E_{rot} \ll E_{vib} \ll E_{el} \,. \tag{2.37}$$

Equation 2.34, describing the total molecular energy, may now be expanded to:

$$E_{tot} = E_{el} + E_{vib} + E_{rot} \,. \tag{2.38}$$

For diatomic and linear polyatomic molecules, rotational motion is restricted to an axis perpendicular to the molecular longitudinal axis. A corresponding diagram is shown in

Figure 2.19. The rotational energy is given by:

$$E_{\text{rot}} = B\,J(J+1) \quad \text{with} \quad J = 0, 1, 2, \ldots, \tag{2.39}$$

where E_{rot} is quantized by the rotational quantum number J. The rotational energy values of a diatomic molecule are therefore restricted to multiples of the rotational constant B. In contrast to vibration, rotational state energies are not equally spaced but show a quadratic increase with J.

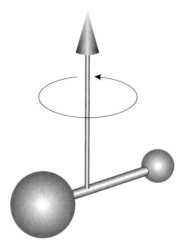

Figure 2.19: Rotational motion of a diatomic molecule

According to Equations 2.37 and 2.38, every vibrational state possesses a substructure, which is composed of a large number of rotational states. On the other hand, the thermal energy of the absorber is high enough to populate many rotational states. Any vibrational band therefore consists of a large number of rotational lines. Figure 2.20 gives an example of such a rotational spectrum within a vibrational band. The spectral distance given for two consecutive rotational lines shows that rotational structure of diatomic molecules may exhibit a strongly modulated absorbance signal within a small spectral interval of the order of 10 pm.

For larger and nonlinear polyatomic molecules, rotational motion may occur with respect to three axes, resulting in a markedly raised number of rotational states. Moreover, the rotational constants determining the energy spacing of the rotational states (refer to Equation 2.39) diminish as a result of increased molecular moments of inertia. Consequently, both the spectral density and quantity of the rotational lines increase. The distance

2. Theoretical Concepts

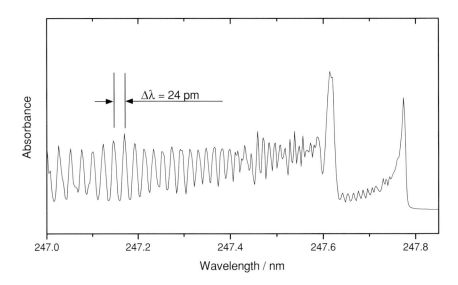

Figure 2.20: Rotational transitions observable as substructure of a vibrational band (PO molecule, refer to Figures 2.16 and 2.18)

between adjacent lines becomes smaller than their widths under conditions of normal pressure and temperature. The modulation of the observable absorbance disappears and degenerates to diffuse, non-structured bands, which are of no effect in HR-CS AAS.

2.3.4 Dissociation Continua

If molecules absorb radiation having quantum energy higher than the dissociation energy, the final state is not a discrete energy state. Above the boundary wavelength one thus observes a spectral continuum without any fine structure. This situation can be compared with the ionization limit in atomic spectra, which corresponds to the energy for the total stripping of a valence electron from the atom (refer to Figure 6.1 in Chapter 6).

Massmann and co-workers [102] undertook detailed studies of these types of spectra in graphite furnaces, and scanned the dissociation continua of numerous molecules, including the alkali halides. As this kind of background absorption is 'continuous', it does not represent any problem for HR-CS AAS, and it can be corrected easily by all BC systems used in LS AAS, unless the background absorption is changing too rapidly or it is exceeding the correction limits discussed earlier. More details may be found in Reference [150].

3. Instrumentation for HR-CS AAS

3.1 Radiation Source

In contrast to a line source (LS) the term continuum source (CS) describes a radiation source, which generates a spectrum with a continuous spectral distribution over a broad wavelength range. Nevertheless a CS can be operated continuously or discontinuously, i.e. in a continuous or in a pulsed mode. An instructive review of CS for common spectroscopic applications has been published by Ingle and Crouch [71].

In principle, incandescent sources, such as quartz-halogen lamps and different types of arc lamps filled with deuterium or xenon may be used for HR-CS AAS. But the most critical parameter for the selection of a radiation source is its spectral radiance in the important far-UV spectral region down to 190 nm. In order to compete with the detection limits of LS AAS the spectral radiance per picometer bandwidth increment of a CS should be at least one order of magnitude higher than the narrow emission lines of hollow cathode lamps, since the geometrical conductance of a HR-CS AAS spectrometer is lower by the same order of magnitude.

Hence, based on the conventional design of a classical xenon short-arc lamp, as used e.g. for stadium illumination, an improved lamp type with increased UV emission has been developed by GLE (Berlin, Germany), and used since for all measurements described in this book (see Figure 3.16 in Section 3.4).

Even though the lamp looks like a conventional arc lamp, it has been optimized to run in the so-called 'hot-spot mode'. This discharge mode is characterized by the appearance of an extremely small plasma spot close to the cathode surface, in contrast to the typical diffuse arc shape of the common xenon lamps (refer to Figure 3.1). The plasma contraction is substantially achieved by selection of sophisticated materials for the anode and cathode rods, a short electrode distance (< 1 mm), an increased xenon pressure (about

3. Instrumentation for HR-CS AAS

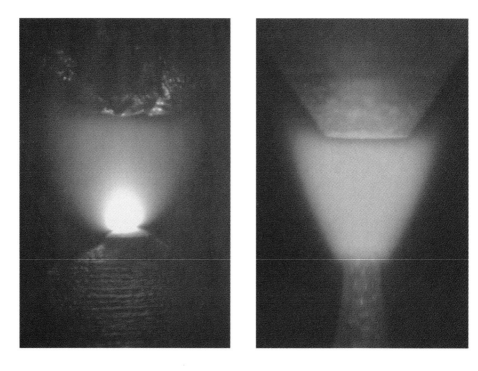

Figure 3.1: Photo of a xenon short-arc lamp operating in hot-spot mode (left) and diffuse mode (right)

17 bar in cold condition), and a specific temperature regime induced by optimized electrode geometries. During operation the inner lamp pressure increases by a factor of 3 – 4 and a hot-spot is formed with a diameter of less than 0.2 mm, and a plasma temperature of about 10 000 K.

The lamp is operated with a nominal power of 300 W (typically 20 V and 15 A) using a DC power supply. An additional circuit to produce a short high-voltage pulse of about 30 kV realizes the ignition. To maintain a tolerable operating temperature, the lamp is mounted in a water-cooled housing, which is integrated in a closed water loop with air cooling.

The wavelength-dependent spectral radiance of the hot-spot xenon lamp is depicted in Figure 3.2 for different distances from the cathode in relation to the prominent emission lines of some hollow cathode lamps. In addition, a comparison of its spectral radiance to a commercial xenon lamp operating in diffuse mode as well as a conventional D_2-lamp is shown in Figure 3.3.

3.1 Radiation Source

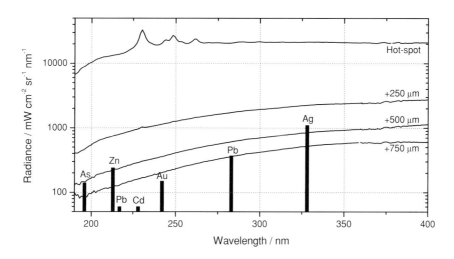

Figure 3.2: Wavelength-dependent spectral radiance of the xenon short-arc lamp measured in the hot-spot and at different distances from the cathode, in comparison to some selected emission lines of hollow cathode lamps

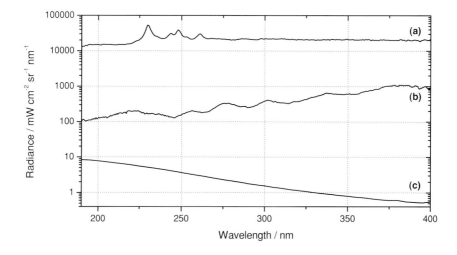

Figure 3.3: Comparison of the wavelength-dependent spectral radiance of (a) the xenon short-arc lamp operating in hot-spot mode (XBO 301, 300 W, GLE Berlin, Germany); (b) a commercial xenon lamp operating in diffuse mode (L 2479, 300 W, Hamamatsu, Japan); (c) a conventional D_2-lamp (MDO 620, 30 W, Heraeus, Germany)

The small size and the typical erratic movement of the hot plasma zone requires a fast active stabilization of the spot image with respect to the spectrometer entrance slit for its optimum illumination. This is realized by a lightweight plane folding mirror close to the lamp, which is controlled by piezo-electrical or magneto-mechanical actuators. This mirror is affected by the signals of a four-quadrant detector, which is irradiated by the lamp using the partial reflection of a thin quartz plate, mounted in front of the spectrometer slit. The illumination control unit runs permanently during the whole AAS recording cycle. In this way, in addition to the correction of the hot-spot jitter, thermal drift, as well as displacement resulting from the replacement of the xenon lamp after some time, are compensated.

The radiance drops constantly over the lifetime of the lamp because of the darkening of the lamp bulb, nevertheless even after some 1000 h of operation it is still well above the 25 % level, which means that the worsening of the LOD, in the case of shot-noise limitation, is still within a factor of two.

3.2 Research Spectrometers with Active Wavelength Stabilization

The requirements of HR-CS AAS on the spectrometer, in contrast to LS AAS, are extremely sophisticated. In particular, the instrumental resolving power has to be about two orders of magnitude higher than in conventional LS AAS spectrometers across the entire AAS spectral range to ensure comparable sensitivity for each element and the greatest possible freedom from spectral interferences due to line overlap. Furthermore only adequate geometrical conductance and diffraction efficiency guarantee the required high SNR to obtain low LOD, primarily in the far-UV range. The truly simultaneous measurement of any analytical line and its spectral vicinity is necessary for accurate background correction. Last but not least, precise absolute wavelength and dispersion control in the femtometer scale is indispensable to guarantee analytical reliability and to take advantage of reference spectra stored in the computer.

The capabilities of classical spectrometers, which generate low order spectra by using either concave or plane gratings, are well known from LS AAS and other analytical techniques [73, 150]. They undoubtedly offer interesting features for a variety of applications characterized particularly by low or medium spectral resolution. But it will be shown in the following sections that only specially designed echelle spectrometers with either internal or external order separation in combination with modern solid-state detectors have the potential to meet all of the requirements listed above for HR-CS AAS.

3.2.1 Echelle Grating

The characteristics and fabrication of echelle (French: ladder) gratings were first described by Harrison in 1949 [56]. Further general information about the theory and application of this grating type is given by Schroeder [127] and Boumans and Vrakking [14].

Due to their coarse ruling structure in the range 30 to 300 grooves per mm, and angles of diffraction greater than 45°, echelle gratings are intermediate in character between classical echellette gratings and Michelson's echelon. Up to now the mechanical ruling of echelle gratings is a complex and time-consuming process and only copies of a limited number of master gratings are available from a few companies.

The shape of an echelle grating and the propagation of the principal rays, which always run close to the auto-collimation mode, are depicted in Figure 3.4. Due to the large step-shaped grooves high order numbers between 20 and 120 are typically generated by an echelle grating. Maximum grating efficiencies up to 70 % and permanent high dispersion are achieved for a wide spectral range by diffracting all overlapping orders near to the high blaze angle perpendicular to the small facet of the step.

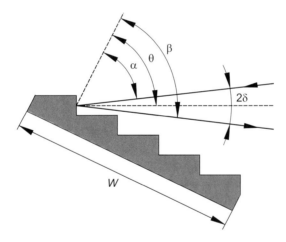

Figure 3.4: Echelle grating (α: angle of incidence, β: angle of diffraction, θ: blaze angle, 2δ: angle between incident and diffracted ray at blaze maximum, W: ruled grating width)

Only two basic equations have to be used to describe the specific features of an echelle grating. At the blaze maximum in each order the angle of incidence α and the angle of diffraction β can be expressed in terms of the small angular difference δ with respect to the blaze angle θ. With $\alpha = \theta + \delta$ and $\beta = \theta - \delta$, as well as $\cos \delta \approx 1$, the universally valid grating equation can be written as:

$$m\lambda = \frac{2W \sin\theta}{N},\qquad(3.1)$$

where m is the order number, λ the wavelength, W the ruled width of the grating and N the total number of grooves. The constant product $m\lambda$ is the basic value for the calculation of the center wavelengths in each echelle order and for the characterization of an echelle spectrometer in general. The difference between the center wavelengths of neighboring orders corresponds to the free spectral range (FSR), and is called $F(\lambda) = \lambda/m$, which is typically in the range of only a few nanometers.

According to the Rayleigh criterion, the theoretical resolution $\Delta\lambda_{theor}$, which could be produced by a spectrometer, is given by the theoretical diffraction-limited resolving power R_{theor}:

$$R_{theor} = \frac{\lambda}{\Delta\lambda_{theor}} = m\,N\,.\qquad(3.2)$$

Using Equations 3.1 and 3.2 the simple formula:

$$R_{theor} = \frac{2W \sin\theta}{\lambda}\qquad(3.3)$$

shows, that the highest resolving power of the echelle is achieved by using a large grating ruled with a great blaze angle. Thus a resolving power of up to several millions is achievable under optimum conditions for large echelles.

A similar formula could be obtained for the angular dispersion $\delta\beta/\delta\lambda$ differentiating the grating Equation 3.1:

$$\frac{\delta\beta}{\delta\lambda} = \frac{2\tan\theta}{\lambda}\,.\qquad(3.4)$$

The angular dispersion of an echelle grating for a given wavelength only depends on its blaze angle. Consequently under normal experimental conditions, where the slit width is considerably greater than the diffraction determined limit, the instrumental resolving power $\lambda/\Delta\lambda_{instr}$ of an echelle spectrometer is a constant per detector pixel at the center wavelength in each order.

Nowadays high-quality echelle gratings are commercially available with blaze angles up to $79\,°$. But even the angular dispersion of a $76\,°$ echelle is about one order of magnitude higher than a classical echellette grating generally used in its first or second order. In this way, high linear dispersion, which is the product of the angular dispersion and the camera focal length, could be obtained at the detector with a very compact echelle spectrometer. This implicates improved thermal and mechanical stability, better transportability and lower manufacturing cost.

According to Equation 3.1 all diffraction orders of the echelle grating superimpose and cannot be initially distinguished by the detector of a spectrometer. Consequently order separation by an additional dispersing component is essential for unambiguous wavelength determination.

The most favored way for a sequential detection of spectral fractions smaller than single echelle orders is the external pre-selection in front of an echelle monochromator, generating a one-dimensional spectrum shape. On the other hand, the simultaneous registration of an enlarged spectral interval requires internal order separation inside an echelle spectrograph equipped with a two-dimensional array detector. Using a cross-dispersing prism a high number of orders covering a broad spectral range can be focused side by side with maximum detector exploitation. It is therefore of particular importance to optimize the size and the quality of the echelle spectrum image by a sophisticated spectrometer design.

3.2.2 Sequential Spectrometer

Various research spectrometers were designed and put into operation for different applications in high-resolution spectroscopy by Becker-Ross and co-workers at the ISAS – Institute for Analytical Sciences (former: Institute of Spectrochemistry and Applied Spectroscopy), Department Berlin, Germany. Beginning in the early 1990s, this unique type of equipment was used systematically for basic investigations in HR-CS AAS. During this period two PhD theses were presented at the ISAS by Weiße [148] and Schütz [128] which addressed selected topics. General considerations and technical features concerning these spectrometers will be described briefly in this and the next section.

Due to the long read-out time and the immense cost of large image detectors the spectrometer design at ISAS was basically focused on the coupling of an echelle monochromator to a linear detection system. Using this concept of a sequential spectrometer, only a relatively small spectral window is detectable simultaneously, but any wavelength position is available within a broad wavelength range. However, the figures of merit concerning resolving power, geometrical conductance, stray light level and spectral accuracy, as well as a short peak hopping time from wavelength to wavelength, are strong arguments in favor of this concept.

A first version of the resulting **D**ouble **E**chelle **Mon**ochromator (DEMON) was already presented in 1993 at the XXVIV CSI in York by Florek et al. [35]. It comprised a prism monochromator creating the order pre-selection in front of a simple echelle monochromator and was used for HR-CS AAS. An improved version of this instrument [58], is shown in Figure 3.5, illustrating the arrangement principle of this concept.

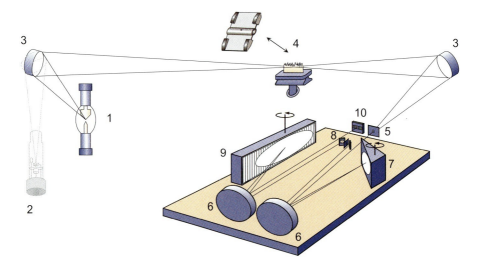

Figure 3.5: HR-CS AAS setup with DEMON spectrometer (1 xenon short-arc lamp, 2 hollow cathode lamp (optional), 3 elliptical mirrors, 4 atomizer, 5 entrance slit, 6 parabolic mirrors, 7 prism, 8 folding mirrors and intermediate slit, 9 echelle grating, 10 CCD detector)

Both DEMON components, the prism and the echelle monochromator, are arranged in a similar Littrow-mounting, coupled by two small folding mirrors. The double-pass mode at the Littrow-prism increases the angular dispersion for minimum prism size and, according to Equations 3.3 and 3.4, the auto-collimation mode of the 76° echelle grating results in maximum dispersion and resolving power. Furthermore a compact design producing minimum aberrations could be realized even for high relative apertures using only one off-axis parabolic mirror as collimator and camera for each monochromator.

The coupling of the high dispersive 76° echelle grating with the powerful low aberration optics results in the enormous instrumental resolving power of 75 000, which corresponds to the FWHM of the instrument profile of 2.7 pm at 200 nm and 6.7 pm at 500 nm, respectively. These values are in good agreement with the spectrometer resolution requirements, derived theoretically in Section 2.2. There it is shown, that the SNR of the absorbance measurement decreases strongly for instrument profiles significantly smaller than twice the absorption profile. For broader instrument profiles the probability of spectral interferences increases, and the reliability of the analytical results suffers when complex chemical matrices are present.

The precise wavelength adjustment is realized by prism and grating rotation actuated by stepper-motor controlled lever arms. All optical and mechanical spectrometer components are arranged on a rugged base made of cast-iron. A photograph of the DEMON spectrometer module is shown in Figure 3.6.

Figure 3.6: Top view of the DEMON research spectrometer (same optical arrangement as shown in Figure 3.5)

The radiation from the CS is focused through the atomizer onto the spectrometer entrance slit using two elliptical mirrors. Herewith the image of the hot-spot is magnified by a factor of six to illuminate the spectrometer with the greatest entrance slit height for maximum geometrical conductance. This results in a relative aperture of about F/1.5 at the lamp, which is, for practical reasons, only manageable using elliptical mirrors.

After the collimated beam is refracted by the prism a small segment of the low-dispersed continuum spectrum passes the intermediate slit, which is, moreover, the entrance slit of the high-resolution echelle monochromator. Rotation of the prism ensures that exactly this spectral interval, which involves the analytical line of interest and its spectral neighborhood, is transmitted. Subsequently the echelle monochromator acts like a magnifying glass and spreads the pre-selected section of one echelle order with substantially higher dispersion.

3. Instrumentation for HR-CS AAS

A trapezoidal intensity profile of the CS spectrum is generated resulting from the convolution process, induced by the small entrance and the wider intermediate slit. At each wavelength the broadness of the intensity profile can be fitted to the detector dimension by motor-controlled adjustment of the intermediate slit width. The adequate profile position on the detector is permanently checked and, when necessary, shifted by prism rotation. Figure 3.7 shows an example of a typical continuum intensity profile. It is sloped due to the echelle blaze profile since the actual diffraction angle differs from the blaze maximum. The contour of the sharp Hg absorption line at 253.652 nm is located in the middle of the profile symmetrical to the center pixel, often surrounded by additional atomic lines or molecular structures. Based on this simultaneously detected spectrum interval the unique HR-CS AAS background correction algorithm is applied as described in detail in Section 5.2.

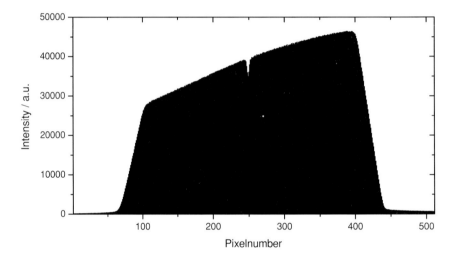

Figure 3.7: Continuum intensity profile with the Hg absorption line at 253.652 nm in the center

Two classes of laboratory spectrometers were assembled according to this concept with substantially different spectral resolution. The *DEMON*, initially designed as a universal research spectrometer, has already been used for several analytical applications [8,151], and is the precursor of a total of five further improved prototypes, following the same instrumental concept. The majority of the basic measurements, the experimental work, resulting in technical improvement of the setup, the testing of the background correction algorithms, and essentially all the analytical evaluation work described in this book have been carried out using these five prototypes (refer to Chapters 4 and 8).

The second one, called *SuperDEMON*, was developed as an instrument with an even higher spectral resolution for fundamental investigations which were necessary to prove theoretical considerations on absorption line profiles and, in particular, on CS AAS calibration curves (refer to Section 4.4). The optical and mechanical design of the SuperDEMON spectrometer is similar to that of the DEMON.

Essentially the SuperDEMON differs mostly in increased resolving power and linear dispersion. To achieve this aim the spectrometer was substantially scaled up regarding all component dimensions and equipped with a high-blaze echelle grating. Additionally an anamorphotic magnification optics consisting of two cylindrical mirrors produces an enlarged image of the primary spectrum on the detector. In this way, as desired, at least three pixels are illuminated for the diffraction limited slit width of about 15 μm.

The indices of the main optical elements together with the specifications of both spectrometer types relevant for methodical development of HR-CS AAS are compiled in Table 3.1.

Table 3.1: Overview of the characteristics of the sequential HR-CS AAS spectrometers

	DEMON	SuperDEMON
echelle grating	75 grooves/mm blaze angle: 76° 270 x 60 mm^2	94.15 grooves/mm blaze angle: 79.9° 360 x 60 mm^2
prism	quartz apex angle: 25°	quartz apex angle: 8°
collimator / camera	relative aperture: F/12 parabolic: $f = 400$ mm off-axis angle: 10°	relative aperture: F/20 parabolic: $f = 1000$ mm off-axis angle: 4°
post-magnification		anamorphotic: 4 - times
wavelength distance per pixel	λ/140 000	λ/1 600 000
instrumental resolving power	75 000	340 000
detector	Hamamatsu S 7031-0906 number of pixels: 512 x 58 pixel size: 24 x 24 μm^2	Hamamatsu S 7031-0906 number of pixels: 1024 x 58 pixel size: 24 x 24 μm^2

3. Instrumentation for HR-CS AAS

The excellent spectral resolution of the SuperDEMON is exhibited in Figure 3.8 by means of the recording of the well-known Fe I quartet at 310 nm emitted by a HCL in comparison to DEMON resolution, which has been optimized for HR-CS AAS routine work.

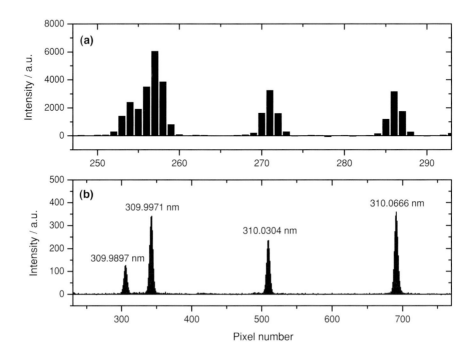

Figure 3.8: Fe I quartet at 310 nm recorded with (a) the DEMON and (b) the SuperDEMON

Nevertheless, only extremely narrow laser lines are applicable to determine the real apparatus function of the SuperDEMON resulting from residual aberrations and detector deficiencies. Figure 3.9 exemplifies the measured emission line profiles of the Nd:YAG laser harmonics at 532 nm and 266 nm, respectively. Their shape looks quite similar but the different angular dispersion given by the echelle grating must be taken into account. These line profiles were fitted applying the well-known Voigt function, and utilized for the deconvolution algorithm described in Section 2.1.5.

Both spectrometers are equipped with a self-controlling wavelength calibration routine using the possibility to avoid the order pre-selection in front of the echelle monochromator. This takes up the well-known method from classical VUV spectrometry, where VUV lines diffracted from echellettes or concave gratings in higher orders were calibrated

3.2 Research Spectrometers with Active Wavelength Stabilization

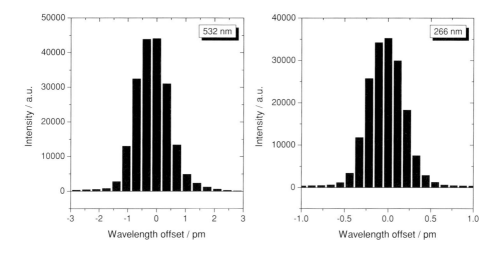

Figure 3.9: SuperDEMON apparatus function for 25 μm entrance slit width determined by means of the Nd:YAG laser harmonics at 532 nm (left) and 266 nm (right)

by adequate first order visible lines. The overlapping of more than hundred orders could be advantageously exploited for echelle spectrometers.

When a number of emission lines from a suitable radiation source passes directly through the intermediate slit, bypassing the prism monochromator, the echelle grating on its own arranges the lines in an apparently confused and mixed assembly. But all lines of overlapping orders, partly far away in wavelength, are imaged close together as a well-defined pattern. Using this principle a simple neon glow discharge lamp can be used for wavelength calibration down to the VUV edge of the AAS wavelength range. Tilting a folding mirror into the optical path, the neon lamp makes available about 30 narrow lines of sufficient intensity from 585 nm to 885 nm (refer to Table 3.2). In Figure 3.10 the odd-looking image of the neon spectrum is depicted in dependence on the angle of diffraction for a 76° echelle grating.

For each grating position at least two lines of the Ne lamp could be focused on the detector simultaneously. In this way, after the initial grating drive running with an accuracy of about 2 arcmin, the wavelength calibration routine simply compares the expected to the measured neon line positions and, when required, initiates a final adjustment by successive grating rotation. At the end of the calibration routine the absorption line of interest is fixed with an absolute accuracy of a few percent of a pixel corresponding to about 0.5 arcsec and the spread of the continuum background spectrum is perfectly adapted to the detector dimension.

3. Instrumentation for HR-CS AAS

Table 3.2: Neon line positions used as reference for the self-controlling wavelength calibration routine

Angle of deflection / °	Wavelength / nm	Grating order	Relative intensity
71.88	616.3594	41	0.050
71.99	702.4050	36	0.016
72.03	588.1895	43	0.191
72.20	703.2413	36	1.000
72.26	602.9997	42	0.024
72.34	633.4428	40	0.120
72.48	724.5167	35	0.761
72.60	650.6528	39	0.277
72.62	667.8277	38	0.219
73.33	653.2882	39	0.138
73.39	849.5360	30	0.096
73.43	621.7281	41	0.081
73.58	607.4338	42	0.111
73.68	671.7043	38	0.197
73.73	638.2991	40	0.407
73.96	594.4834	43	0.104
74.27	609.6163	42	0.205
74.31	640.2246	40	0.382
74.41	753.5774	34	0.057
74.55	692.9467	37	0.343
74.63	754.4046	36	0.024
74.84	885.3866	29	0.026
74.97	626.6495	41	0.221
74.99	597.5534	43	0.034
75.32	659.8953	39	0.151
75.45	585.2448	44	0.556
75.87	614.3063	42	0.261
76.08	717.3938	36	0.048
76.29	630.4789	41	0.075

3.2 Research Spectrometers with Active Wavelength Stabilization

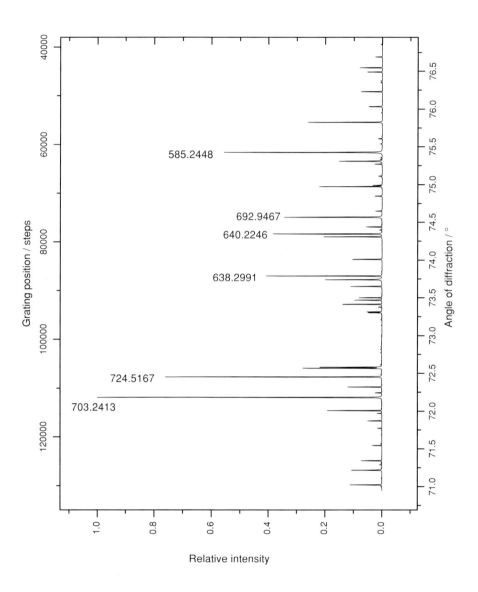

Figure 3.10: Neon line positions in dependence on the angle of diffraction and corresponding grating positions

3.2.3 Simultaneous Spectrometer

To make available the maximum spectral information for HR-CS AAS methodical development, such as the selection of alternative analytical lines and the characterization of spectral interferences, it was necessary to measure accurately overview absorbance spectra of relevant elements (refer to Chapter 6) as well as of diatomic molecules which show not only broad-band absorption, but also fine structures including a large number of extremely narrow lines (refer to Chapter 7). A scanning procedure using an echelle monochromator would obviously be a quite ineffective method for generating these large data files. Thus the laboratory setup of an echelle spectrograph formerly designed by Florek et al. was applied for simultaneous registration of broad-range emission spectra [36].

The design of the **Ar**ray **E**chelle **S**pectrograph (ARES) is characterized by high spectroscopic performance resulting from effective mechanisms for aberration correction and active wavelength stabilization. This is achieved by arranging the echelle grating in the so-called 'tetrahedral mounting', indicated by the fact that collimator and camera mirror on one side as well as entrance slit and detector on the other side describe the corners of a tetrahedron (refer to Figure 3.11). The planes of reflection at both mirrors are perpendicular to the plane of the echelle diffraction. In this way, the compensation of the prominent aberration term coma could be realized for minimum deviation angles at the concave mirrors according to the geometric conditions already described by Czerny and Turner [23].

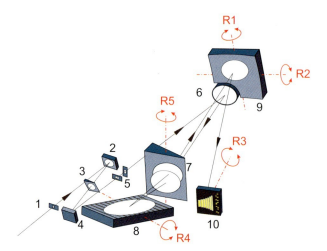

Figure 3.11: Schematic setup of the ARES spectrograph; (1) entrance slit, (2–5) dispersive slit illumination (DSI), (6, 9) spherical mirrors, (7) prism, (8) echelle grating, (10) CCD array detector, (R1–R5) piezo-electrically controlled rotation units

Due to the fact that the spectrometer optics utilize only spherical collimator and camera mirrors, the resulting astigmatism term is compensated by the separation of the entrance slit into a horizontal (slit width) and a vertical (slit height) component. Each slit component is located at a different distance from the collimator. In this way the systematically 'introduced' astigmatism is exactly compensated by the spectrograph optics and order overlap caused by astigmatically elongated slit images is avoided.

The resulting excellent image quality extending over the large spectrum area with dimensions of about a tenth part of the camera focal length in both directions is illustrated in Figure 3.12.

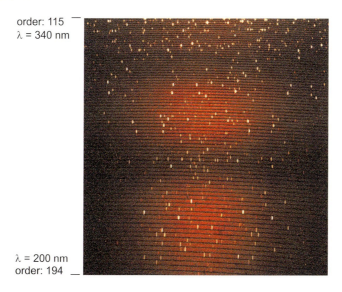

order: 115
λ = 340 nm

λ = 200 nm
order: 194

Figure 3.12: Echelle spectrum section of a platinum HCL mixed with D_2 continuum in the wavelength range from 200 nm to 340 nm

The internal order separation is favorably realized by a quartz prism in double pass mode in front of the echelle grating. In contrast to the alternative application of a cross-dispersing grating, only a prism guarantees high transmittance over the wide spectral range and a unique wavelength separation free from additional order overlap. But, consequently, the lateral distances between adjacent echelle orders of the two-dimensional spectrum image decrease with increasing wavelengths resulting from the interaction of prism cross-dispersion and echelle FSR. To ensure clear order separation within the complete spectrum the fixed spectrograph slit height is stringently limited according to the minimum spacing which appears between the long wavelength orders. Consequently the geometrical conductance of echelle spectrographs, which is generally moderate under the condition of high resolution, is further reduced. However, it can be substantially enhanced in the UV by coupling the ARES with the **D**ispersive **S**lit **I**llumination (DSI) as shown in Figure 3.11.

The DSI produces an astigmatic image of a 'classical' entrance slit and perfectly illuminates the separated slit components, which are essential for the tetrahedral mounting of the echelle grating. Additionally, it creates a very low dispersion spectrum of the radiation source on the vertical slit component resulting from the insignificant prism. Now both the entrance slit height and the width of the vertical slit component can be expanded to the value of maximum order distance in the UV part of the echelle spectrum. The appropriate adjustment always lets the shortest wavelength beam pass the wide slits completely. All other beams are partially vignetted. Due to this vignetting effect of the vertical slit component the effective slit height depends on the wavelength and can be well adapted to the non-uniform order spacing. Constant and narrow non-illuminated interspaces between neighboring echelle orders appear and nearly full spatial exploitation of the detector area is achieved (see Figure 3.12). By running a dedicated binning procedure, which sums the intensities of pixels corresponding to equal wavelengths, the SNR in the VUV range is significantly increased. The main characteristics of the ARES and DSI components are summarized in Table 3.3.

Table 3.3: Overview of the characteristics of the ARES and DSI components

	DSI	ARES
echelle grating		50 grooves/mm blaze angle: 76° 120 x 30 mm²
prism	quartz apex angle: 3°	quartz apex angle: 25°
collimator	relative aperture: F/10 spherical: $R = 200$ mm	relative aperture: F/10 spherical: $R = 500$ mm
camera	spherical: $R = 200$ mm	spherical: $R = 500$ mm
wavelength distance per pixel		$\lambda/80\,000$
detector		Kodak KAF 1000 CCD number of pixels: 1024 x 1024 pixel size: 24 x 24 μm²

3.2 Research Spectrometers with Active Wavelength Stabilization

Using the large blaze angle of the echelle grating a high linear dispersion is achieved for a relatively short camera focal length of only 0.25 m. This results in a very compact and stable spectrograph construction, which is shown in Figure 3.13.

Figure 3.13: Top view of the ARES research spectrometer

The ARES spectrograph is also equipped with an active wavelength stabilization system, which guarantees fixed shape and position of the echelle spectrum pattern relative to the detector. Altogether five micro-positioning units based on piezo-electrical actuators (four active; R1 – R4 and one passive thermo-mechanical R5) are implemented to control permanently the relative spectrum geometry (see Figure 3.11).

3. Instrumentation for HR-CS AAS

The actuators affect directly the echelle diffraction, the prism dispersion, two lateral shift directions and the spectrum rotation. Each read-out of the detector gives the full information about the current spectrum position and geometry by evaluating the centers of gravity of at least three emission or absorption line positions. Preferably the positions of one arbitrary line appearing two-fold at the ends of adjacent low orders in the long wavelength range and one additional line at the short wavelength edge of the echelle pattern are used for sensitive determination of the spectrum status. A sophisticated on-line correction algorithm calculates the set values for the corresponding piezo-actuators after analyzing the differences between the actual and the reference line coordinates. If required the micro-positioning units are adjusted in a hierarchic sequence.

Following this concept, first one of the reference lines in the long wavelength range is focused exactly on the corresponding target pixel by rotation of the camera mirror around two perpendicular axes (R1, R2). Then the distance between the two low order lines indicates the status of the linear dispersion, which is influenced mostly by thermal expansion of the grating and the detector. Correction of the linear dispersion is feasible by variation of the angle of incidence at the echelle grating. This is achieved by rotating the grating around an axis parallel to its grooves (R4). Finally the rotation of the detector around an axis perpendicular to its surface is used to achieve the best fit between the order orientation and the detector pixel pattern (R3). The only passive actuator (R5) affects the prism tilt, similar to the mechanism of a thermally-compensated clock pendulum, and stabilizes the temperature-dependent prism dispersion by changing the angle of radiation incidence.

After running this adjustment routine the deviation of the real wavelength scale from the absolute one is of the order of $\lambda/8\,000\,000$ without any thermostatic control of the instrument. More detailed information about the principle and the figures of merit of the active wavelength stabilization is given in a publication of Becker-Ross et al. [9].

3.3 Detector

The requirements on the radiation detector of the high-resolution spectrometer are very complex in order to exploit fully the methodical potential of HR-CS AAS. This barrier already exists for the design of a fast sequential system using flame atomization, but to an even greater extent for a simultaneous broad-range setup for a graphite furnace atomizer. Some needs are already fulfilled, other criteria are still critical or subject to trade-offs and, finally, improved detector parameters are desirable. In detail, the CS AAS detector has to offer high quantum efficiency from the far-UV to the NIR, fast and low-noise read-out, high saturation capacity, as well as high spatial resolution for arbitrary two-dimensional pixel numbers and last, but not least, moderate cost. Hence, the history of CS AAS development is strongly influenced by the progress in detector technology.

3.3 Detector

In the very beginning of CS AAS work the data generation had started by using a real 'multichannel' photon detector – the photoplate. This combined advantageously high spatial resolution and broad-range spectral sensitivity. However, this early image detector fell into oblivion, because it suffered from poor dynamic range and low repetition rate due to the chemical developing process of the plate and time-consuming spectrum evaluation.

The photo-multiplier tube (PMT) as a one-channel detector could not be an alternative. Even though the PMT is an almost perfect photon detector, offering shot-noise limited measurements for almost every photon flux, but its restriction to exclusive sequential registration of spectral intervals makes it, and also the different types of dissector tubes, unsuitable for application in CS AAS.

With the commercial availability of solid-state array detectors the advantages of both the photoplate and the PMT could be merged and a sort of 'electronic' photoplate was created. Especially, photodiode arrays (PDA) and charge-coupled devices (CCD) offer the effective registration of spatially and temporally resolved intensity distributions, as has been discussed in detail by Harnly and Fields [53]. Nowadays, the different layouts of these array detectors, characterized by extremely high mechanical accuracy and photometric performance, in combination with the permanently improving potential of electronic signal processing, permit the design of HR-CS AAS spectrometers for routine analytical work. In this section some detector considerations are discussed, which were of some relevance for the design of the different research spectrometers mentioned in this book.

The performance of each array detector, not only for CS AAS, is described in particular by its 'dynamic range', which means the application area of the detector, where shot-noise limited absorbance measurements are possible. It depends on basic detector parameters and is calculated by the ratio between the saturation capacity and the square of the read-out noise.

The importance of a large dynamic range becomes obvious when taking into account measurements with strong background absorption as well as simultaneous measurements over a wide spectral region. From this point of view, CCD arrays are the most favored CS AAS detectors and were selected for the research spectrometers. Typical values for the saturation capacity between 600 000 and 800 000 electrons per pixel and a read-out noise of 5 to 30 electrons result in a shot-noise limited dynamic range of 600 to 800, which is quite sufficient for AAS applications.

The implementation of so-called 'back-thinned' CCD removes sensitivity limitations of classical front-illuminated and UV-coated CCD. Using this improved technology, the substrates with extremely thin silica layers are back-illuminated and a large increase in quantum efficiency in the UV range up to 0.9 electrons per photon can be achieved. Specific anti-reflection layers on the detector surface produce further enhancement of the detector efficiency for certain spectral bands.

3. Instrumentation for HR-CS AAS

The read-out regime for a CCD array detector, when it is limited to a small number of pixels (m) perpendicular to the read-out direction, is truly simultaneous. The electrons, which are generated in each pixel well during illumination, are shifted simultaneously and very rapidly to the read-out register after each accumulation interval (refer to Figure 3.14). In this way, using the so-called 'binning' procedure, all electrons within a column, which represent the same wavelength, are collected in one register increment before reading via an amplifier. Hence, n small linear spectral intervals can still be detected by a CCD spectrometer for substantial slit heights without a mechanical shutter. This specific attribute can also be exploited advantageously for echelle spectrometers of the DEMON type (see Section 3.2.2).

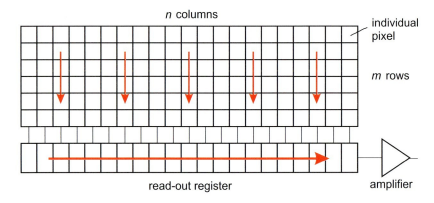

Figure 3.14: Read-out scheme of a CCD array detector

However, for echelle spectrographs of the ARES type, large CCD detectors with millions of pixels have to be used in order to record the broad spectral interval during each read-out cycle (refer to Section 3.2.3). Unfortunately, to reduce the read-out time and to achieve higher SNR, the application of smart 'on-chip' binning routines is not practicable in this case of simultaneous HR-CS AAS, because of the fact that the variable tilt of successive echelle orders, which is not matched to the rectangular pixel pattern of the detector, would then produce substantial cross-talk between the detector areas to be integrated before A/D conversion. Thus, the individual read-out of all pixels is unavoidable and results in a slow repetition rate whenever a high dynamic range is required.

In general this is not a crucial limitation for basic investigations especially in simultaneous HR-CS F AAS, as it will be shown in Chapter 7. But the question of the appropriate detector is still open to design a sufficiently qualified instrument, which offers specifications comparable to competitive analytical methods, a topic that will be discussed in Chapter 9.

3.4 The contrAA 300 from Analytik Jena AG

The atomic absorption spectrometer contrAA 300 from Analytik Jena AG (Jena, Germany) is the first commercially available instrument for HR-CS AAS, and it is closely related to the DEMON research spectrometers described in detail in Section 3.2.2. At the time of writing this book, the instrument was available only for flame atomization and chemical vapor generation techniques (refer to Figure 3.15). However, the manufacturer has already announced that a combined flame and graphite furnace version will be introduced in the near future.

Figure 3.15: ContrAA 300 – the first commercially available continuum source flame AA spectrometer (Analytik Jena AG, Jena, Germany)

The DEMON design has been adopted for the double monochromator, with a prism pre-monochromator and an echelle monochromator for high resolution of the selected spectral interval, with only minor modification of the optical and analytical specifications. The usable wavelength range extends from 189 nm in the beginning VUV to 900 nm in the near-IR, and covers completely that of LS AAS. The wavelength distance per pixel of $\lambda/140\,000$ is also identical to that of the latest prototypes of the DEMON spectrometers.

3. Instrumentation for HR-CS AAS

The CCD detector has been developed in close cooperation with Hamamatsu (Hamamatsu, Japan), and adapted for the special requirements of HR-CS AAS. A total of 576 pixels that can be used for measurement, make possible the highly-resolved and truly-simultaneous observation of a spectral range of almost 1 nm. The wavelength change takes about 2 s, including all correction and optimization steps. The vertical arrangement of the spectrometer makes possible a very compact design of the instrument of only 781 x 730 x 635 mm^3 (width x depth x height), which is shown in Figure 3.15. The weight of the instrument is around 85 kg.

The high-pressure xenon short-arc lamp, which is depicted in Figure 3.16, is mounted in a water-cooled brass housing. The radiation source is pre-adjusted in its housing, and the whole unit is self-adjusting when the radiation source has to be exchanged. The water cooling of the lamp is an integral part of the instrument. Once mounted, the radiation source is essentially maintenance-free during its specified lifetime, and is operational immediately after ignition.

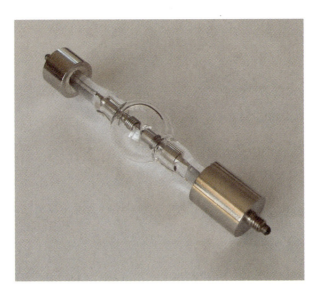

Figure 3.16: Photo of the high-pressure xenon short-arc lamp (GLE, Berlin, Germany)

Atomizer unit, gas control, automatic burner adjustment and sample introduction systems, such as micro-sample injection and automatic sample dilution have been adapted from the proven state-of-the-art LS AAS systems. The working conditions of HR-CS AAS regarding temperature range, gas supply and electrical conditions are essentially as undemanding as for classical LS AAS.

3.4 The contrAA 300 from Analytik Jena AG

The third unit, besides the radiation source and the spectrometer, including the detector, is the instrument control including the operator surface, which is shown in Figure 3.17. The optimum measurement conditions are checked for each spectral range and adjusted before each measurement within a few milliseconds. When the full capability of the system is utilized, a complete set of data from 576 simultaneously read pixels is obtained per millisecond during data evaluation. The data are pre-averaged and the averaged data are transmitted to the application software.

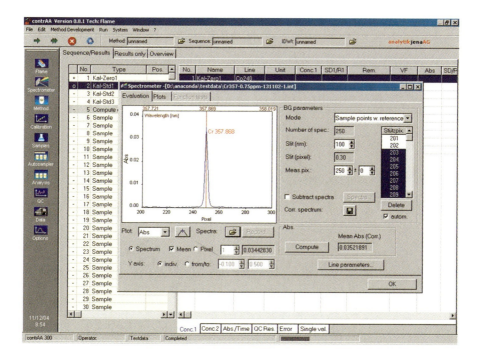

Figure 3.17: Screen-shot of the operator surface of the contrAA 300 spectrometer

The application software is programmed for the manual as well as for the fully automatic measurement mode and supports the analyst in the development and optimization of methods for HR-CS AAS. In addition to the well-known software options of LS AAS, the fast wavelength/element change for flame operation offers a rapid sequential multi-element determination for up to 20 arbitrarily selectable elements using a constant sample flow. The spectral environment can be visualized and evaluated by the analyst during the measurement cycle.

55

Reference spectra are recorded at regular intervals in order to calculate the absorbance values. The purpose of the following data handling is to eliminate fluctuations in the intensity of the radiation source, and to identify appropriate pixels for the simultaneous BC. The algorithm for the automatic baseline adaptation produces reliable results, even without programming by the analyst. Optionally, the spectral range as well as analyte and correction pixels can be defined by the analyst manually. For that purpose the software offers the re-processing of recorded data. Additional algorithms, based on least-squares adaptation, are available for complex analytical problems with the goal being to eliminate well-known spectral interferences using reference spectra for correction.

4. Special Features of HR-CS AAS

4.1 The Modulation Principle

Talking about his early thoughts about AAS back in 1952, Walsh [147] remembers that '... for several years prior to these first thoughts on atomic absorption, I had been regularly using a commercial IR spectrophotometer employing a modulated light source and synchronously tuned detection system. A feature of this system is that any radiation emitted by the sample produces no signal at the output of the detection system. This experience had no doubt prevented the formation of any possible mental block associated with absorption measurements on luminous atomic vapors.' This modulation principle, using either an AC-operated radiation source or a chopper in the radiation beam, and a selective amplifier tuned to the same modulation frequency, has ever since been applied in all commercially available atomic absorption spectrometers. It has been considered one of the major advantages of AAS compared to optical emission techniques, guaranteeing, together with the element-specific radiation source, the selectivity and specificity of atomic absorption measurements.

However, such 'luminous atomic vapors' or any other emission from the atomizer, such as that from the red-hot graphite tube, can cause a problem in AAS only when they are within the spectral bandwidth that is used for measurement, and when their intensity becomes significant in comparison to that of the primary radiation source. Both of these conditions are not met in HR-CS AAS. Firstly, the intensity of the radiation source used for the latter technique is about 1–2 orders of magnitude higher than that of a conventional line source for LS AAS, decreasing the risk for emission interferences by the same magnitude. Secondly, as the detectors used for LS AAS integrate the radiation over the spectral bandwidth, i.e. typically over several 100 pm, any atomic emission within this spectral range would be recorded and amplified. The situation might be even worse for broadband emission, which is integrated over the entire bandwidth, making it some two orders of magnitude greater than on the analytical wavelength. In HR-CS AAS, any atomic emission that does not directly coincide with the analytical line is not measured by the analytical pixel(s), and the intensity of any broadband emission is some two orders of magnitude lower on the analytical pixel(s) compared to its integrated intensity in LS AAS. This means that it becomes negligible, particularly when the much higher lamp intensity is considered as well. Last but not least, in the very unlikely case that there is some extremely intense emission from the atomizer that contributes to the measured signal, the background correction systems, which will be discussed in detail in Section 5.2, would compensate for this potential interference as well.

This means that HR-CS AAS, due to its special features, does not need any modulation of the source or any selective amplifier. This also means that a potential source of noise has been eliminated, as both AC operation of hollow cathode lamps and mechanical choppers contribute to noise in LS AAS. In addition, other problems that are associated with strong emission of the atomizer source in LS AAS, such as the 'emission noise' caused by the nitrous oxide / acetylene flame in the determination of barium and calcium due to the CN band emission, are equally absent in CS AAS for the same reasons, i.e. the much higher intensity of the primary radiation source, and the high resolution.

4.2 Simultaneous Double-beam Concept

In LS AAS it is normal to distinguish between single-beam and double-beam spectrometers. In a single-beam spectrometer the primary radiation is conducted through the absorption volume without geometric beam splitting, while in a double-beam spectrometer the beam is divided into two sections. A portion of the radiation, the *sampling radiation* (or sample beam), is passed through the absorption volume (flame, furnace etc.), while the other portion, the *reference radiation* (or reference beam), bypasses the absorption volume. The geometric splitting and recombining of the beam can be realized by means of a rotating chopper made from a partially mirrored rotating quartz disk, by means of semi-transparent mirrors, or by combinations of these. Typical switching (modulation) frequencies are between 50 Hz and 100 Hz.

Single-beam spectrometers have the advantage that they contain fewer optical components and thus radiation losses are lower and the effective optical conductance higher. The major advantage claimed for *double-beam spectrometers* is better long-term stability since they compensate for changes in the intensity of the source and the sensitivity of the detector. Nevertheless, this advantage is frequently overvalued, since during the warm-up phase not only the radiant intensity of the source changes, but also the line profile, and thereby the sensitivity. The double-beam system also is neither capable of recognizing nor of compensating for changes in the atomizer, such as flame drift during the warm-up phase of the burner. Nowadays it has been kind of accepted that a double-beam spectrometer has clear advantages for routine F AAS in order to avoid frequent checks for baseline drift and recalibration. For GF AAS, where usually the baseline is re-set before each atomization stage, long-term stability is not of interest so that the better optical conductance of a single-beam spectrometer is of advantage.

All spectrometers for HR-CS AAS are optical single-beam instruments, as there is no geometric splitting and recombining of the beam included. However, HR-CS AAS actually offers a far superior stabilization system that can take over all the functions of an optical double-beam instrument with the great advantage of working simultaneously, and not sequentially, as optical systems do. The CCD array detector of a HR-CS AA spectrometer typically consists of several hundred pixels, and they all have their individual charge-collecting capacity, i.e. they can be considered as independent detectors, which are all illuminated by the same radiation source, and which are all read out simultaneously. As only a few of these pixels are typically used to measure atomic absorption (refer to Section 4.4), any other pixel or set of pixels may be used to correct for fluctuations in the lamp intensity, as these fluctuations in a vicinity of some hundred pm are obviously the same for all pixels. And as all pixels are illuminated and read out simultaneously, and not sequentially, even the fastest changes in emission intensity will be corrected perfectly, as shown in Figure 4.1.

Moreover, as the entire radiation passes through the atomizer, any temporal change in the transmittance, for example of the flame gases, is corrected in the same way, a feature that is obviously not available in an optical double-beam system. This function is of particular importance for the measurement and correction of background absorption, which will be treated in detail in Section 5.2.

4.3 Selection of Analytical Lines

Usually, in LS AAS, the most sensitive analytical line is used for the determination of an element, because AAS is mostly applied for trace and ultra-trace analysis, which obviously requires the highest sensitivity. Another reason for using the most sensitive line is that it makes it possible to apply higher dilution in the case of complex sample matrices, and hence to avoid potential interferences. On occasions, however, the most sensitive line is not recommended in LS AAS, as it does not provide the best SNR, as in the case of the 217.001 nm lead line. Another reason not to recommend use of the most sensitive line might be a strongly non-linear working curve due to the presence of other lines in the lamp spectrum that cannot be excluded even with a 0.2 nm bandwidth, as in the case of the 340.725 nm cobalt, the 232.003 nm nickel, or the 244.791 nm palladium line [150].

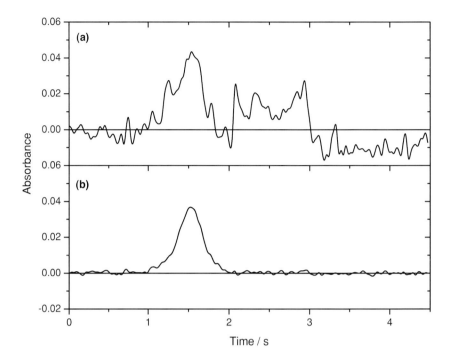

Figure 4.1: Absorbance over time for 200 pg of lead in aqueous solution measured at the CP at 283.306 nm only, (a) without and (b) with the use of reference pixels to correct for lamp noise

All these limitations do not exist in HR-CS AAS, firstly because the radiation intensity of the source is, even down to 200 nm, high enough to provide a significantly better SNR than the line sources in conventional LS AAS. In a first approximation the radiation intensity, and hence the SNR for all lines is of the same magnitude, although it degrades somewhat in the far-UV. Secondly, the resolution of the monochromator is such that only about two times the FWHM of the line, or less, is detected by the analytical pixel. As the source emits a continuum, and the analytical pixel is adjusted in such a way that it is always in the line center, all the phenomena that are associated with a line source, such as line shift, self-absorption or the presence of other lines emitted by the lamp, are non-existent.

Although AAS is predominantly used as a technique for trace analysis, it has also been applied for the determination of main components, such as calcium in cement [21], lead and tin in soft solders [149], or chromium, iron and nickel in stainless steel [16] using flame AAS, to give just a few examples. It has been shown in these applications that the necessary precision cannot be obtained by diluting the sample solution until the analyte content comes within the working range of the most sensitive line. The only successful approach for this application is the use of less sensitive, secondary lines, and only a moderate dilution of the sample solution, as the imprecision introduced by excessive dilution often impairs the results to a degree that is no longer acceptable in this kind of analysis. On the other hand, however, the number of analytical lines with a suitable low sensitivity that could be applied for this purpose is often very limited because of the weak emission intensity provided by the HCL, which does not provide the SNR necessary for high-precision determinations.

In GF AAS with solutions the use of secondary lines is much less common, as high analyte concentrations are much more easily and more precisely determined by F AAS. However, a significant demand for a reduction of sensitivity has been created by the technique of direct solid sample analysis by GF AAS (refer to Section 8.2.2). As solid samples cannot be easily 'diluted', although graphite powder has been used for this purpose, and as the sample mass can only be varied within about one order of magnitude without deteriorating precision, the use of alternate, less sensitive lines is the most obvious way to reduce sensitivity for the determination of higher concentrations. This demand for alternate lines will increase with HR-CS AAS, as this technique has already been shown to be particularly suited to the direct analysis of solid samples, as will be shown for several examples in Section 8.2. In LS AAS the use of secondary lines has often been discouraged, mostly for two reasons. Firstly, secondary lines were often 'weak' lines, i.e. the emission intensity was low, compared to the primary line(s), resulting in a deterioration of the SNR, and hence the precision of the measurement. Secondly, these lines were often poorly investigated, and little was known about potential spectral interferences when these lines were used. Hence often a general warning was issued to be particularly careful with spectral interferences when using secondary lines.

In HR-CS AAS again these problems are essentially non-existent for the same reasons as given above. Firstly, because of the relatively constant, very intense emission of the primary radiation source, there are no weak lines, i.e. the same high SNR ratio will be obtained on all analytical lines, independent of their spectral origin. The only criteria that will have an influence will be the absorption coefficient and the population of the low excitation level when non-resonance lines are used. Secondly, because of the high resolution of the monochromator, and as the entire spectral environment of the analytical line becomes visible in HR-CS AAS, potential spectral interferences can easily be detected, and in addition cannot influence the actual measurement, except in the rare case of direct line overlap. But even in this case HR-CS AAS provides an appropriate solution, as will be discussed in Section 5.2.3.

4.4 Sensitivity and Working Range

The relationship between the concentration c or the mass m of the analyte in the measurement solution or measurement portion as the desired quantity, and the absorbance A or integrated absorbance A_int as the measured quantity is established and described mathematically in AAS by the use of calibration samples, usually calibration solutions. The relationship between the absorbance or the integrated absorbance and the analyte concentration or mass is given by the calibration function, e.g.:

$$A = f(c) \quad \text{or} \quad A_\text{int} = f'(m) . \tag{4.1}$$

The slope S of the calibration function is termed sensitivity, i.e.:

$$S = \frac{\delta A}{\delta c} \quad \text{or} \quad S' = \frac{\delta A_\text{int}}{\delta m} . \tag{4.2}$$

To provide a measure of the sensitivity of the analyte under given conditions one can use the terms 'characteristic concentration' c_0, and 'characteristic mass' m_0 in AAS. This is the concentration or mass of analyte corresponding to an absorbance $A = 0.0044$ (1 % absorption) or an integrated absorbance $A_\text{int} = 0.0044$ s. The sensitivity in AAS is thus given by physical quantities, such as the absorption coefficient of the analytical line (refer to Section 2.1) and also by characteristics of the atomizer unit. The characteristic mass m_0 may be calculated according to:

$$m_0 \, [\text{pg}] = \frac{0.0044 \cdot c \, [\mu\text{g/L}] \cdot V \, [\mu\text{L}]}{A_\text{Standard} - A_\text{Blank}} , \tag{4.3}$$

where c and V are the concentration and the volume of standard required for an integrated absorbance of A_Standard, and A_Blank is the absorbance signal for the blank.

In order to better compare the performances of various Echelle monochromator and slit configurations Smith and Harnly [136] introduced the term 'intrinsic mass'. After the fashion of characteristic mass, intrinsic mass is the analyte mass required to produce an integrated absorbance of 0.0044 s per picometer of spectral bandwidth, calculated from the wavelength-integrated absorbance A_λ (over n_sam pixels) and the spectral bandwidth $\Delta\lambda$:

$$A_\lambda = \Delta\lambda \sum_i^{n_\text{sam}} \log \frac{I_0}{I_i} , \tag{4.4}$$

whereby I_0 is the reference intensity and I_i the intensity of the individual pixel.

In HR-CS AAS, the best linearity of the calibration function is obtained when the absorbance measured at the center pixel (CP), i.e. at the line core only, is used for evaluation. However, the sensitivity might be increased by typically a factor of two by integrating the absorbance over the CP ± 1, i.e. by including part of the line wings in the measurement, as will be discussed in detail later in this section.

A major detraction of LS AAS has always been the relatively short linear region of the calibration curves, from two to three orders of magnitude of concentration. The limits of the linear working range arise from stray radiation and the finite width of the emission lines of the radiation source, which is not monochromatic and just 3–5 times narrower than the absorption profile. With CS AAS, there is no theoretical limit to the calibration range, only the practical limits imposed by the size of the array detector, the increasing possibility of spectral interferences and the ability to clean the atomizer after extremely high analyte concentrations have been introduced [54]. Harnly and co-workers [51, 157] proposed two different approaches to extend the calibration range of CS AAS to some 5–6 orders of magnitude in concentration, the wavelength integrated absorbance (WIA) approach for the read-noise limited linear photodiode array (PDA) detector, and the wavelength selected absorbance (WSA) approach for the shot-noise limited CCD detector.

In the case of the WIA approach, large entrance slits are used and the absorbance is integrated over an increasing number of pixels to measure the integrated absorbance. The shape of the calibration curves obtained with this approach, which is similar to that of the curves for large quotients between the width of the instrumental profile and the absorption line FWHM (refer to Figure 2.13), has been predicted theoretically already by Mitchell and Zemansky [105]. At low concentrations, absorbance increases linearly with increase in concentration. This relationship provides linear plots with a slope of 1.0 when the logarithm of absorbance is plotted versus the logarithm of concentration. As the concentration increases, the absorbance at the peak center reaches a maximum, determined by the stray radiation. At this point, the calibration curves for LS AAS will reach a plateau. With a continuum source, however, absorption in the wings of the profile can be measured. At any wavelength in the wings, absorbance will increase linearly with increase in concentration. However, the wings broaden as a function of the square root of the concentration, hence the wavelength integrated absorbance will also increase with the square root of the concentration. This relationship at high concentrations provides a linear plot with a slope of 0.5 on a log-log plot, as can be seen in Figure 2.13.

The inflection point, the point where the calibration curve makes the transition from a slope of 1.0 to a slope of 0.5, is determined by the 'α-value' and the hyperfine splitting of the absorption profile. The α-value is proportional to the ratio of the *collisional* width to the *Doppler* width and determines the width of the absorption profile. It will vary between elements and for each element as a function of temperature. Hyperfine splitting is determined by the coupling of the electronic transitions and determines the number of components of the absorption profile. Elements with low α-values and fewer hyperfine components will have deeper, narrower profiles and elements with larger α-values and more hyperfine components will have shallower, broader profiles. The former elements will reach the stray radiation limit sooner and have inflection points at lower concentrations [54].

The usefulness of this approach for practical analytical work, however, although it is certainly correct, has to be questioned for several reasons. Firstly, log-log calibration curves are not used in normal analytical work; secondly, the inflection point, i.e. the transition between two linear regions will for sure cause problems for the integration of dynamic signals, as they are produced in GF AAS, as has been criticized correctly by L'vov [99]; and thirdly, the risk for spectral interferences is growing dramatically when some 30 pixels are used for signal evaluation. Anyway, this WIA approach has been proposed for PDA detectors, which are not ideal for CS AAS, and which are not used in the instruments discussed in this book.

The second, the WSA approach proposed by Harnly et al. [50, 51] for use with CCD detectors, in contrast, appears to be much more practical. In this case narrower entrance slits may be used and, in addition to the evaluation of the CP only, i.e. the line core (or the CP ± 1 or ± 2 etc.), other pairs of pixels, such as $+3$ and -3 or $+4$ and -4 are used for evaluation. These latter pixels measure the absorbance at the wings only, at a given distance from the line center, where the absorbance is still increasing linearly with the concentration. In this case the measurement is not influenced by line broadening, as the absorbance around the line center is not considered. This means that a set of linear calibration curves with a slope of 1.0 on a log-log scale is obtained.

Heitmann et al. [62] have demonstrated the performance of a HR-CS AAS instrument for two elements, silver at 328.068 nm with a very narrow absorption line, which is barely resolved by the spectrometer, and indium at 303.936 nm, which extends clearly over more than 5 pixels, as shown in Figure 4.2.

As mentioned above, using HR-CS AAS, signal registration cannot only be done at the CP **M**, corresponding to the line core, as in conventional LS AAS, but also using the volume $\mathbf{M}_x$ of the absorption peak with an increasing number of pixels, or the side pixels $\mathbf{N}_x$ only, as shown in Figure 4.3.

In the first case of peak volume registration, i.e. when an increasing number of pixels is used for detection, the sensitivity can be improved, depending on the peak width. In the case of a narrow absorption line, as is shown for the example of silver, the sensitivity obtained for the CP only is comparable with that obtained with a HCL, as shown in Figure 4.4 (a). Using three pixels, i.e. CP ± 1, results in a significant increase in sensitivity and in the linear working range, but a further extension of the peak volume detection does not make sense, as the signal drops rapidly, and only the noise would be integrated. The situation changes in the case of a broader absorption line, as shown in Figure 4.4 (b) for the case of indium, where the sensitivity obtained in HR-CS AAS is significantly lower than that in LS AAS when the CP only is used for measurement.

4.4 Sensitivity and Working Range

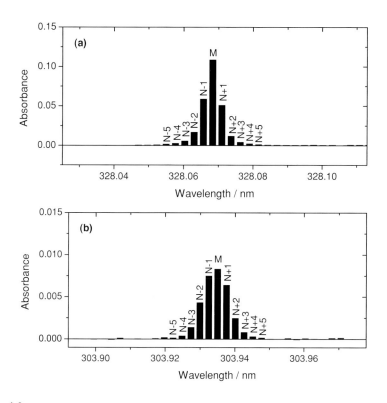

Figure 4.2: Typical line profiles of (a) silver at 328.068 nm and (b) indium at 303.936 nm; sample concentration: 1 mg/L, respectively

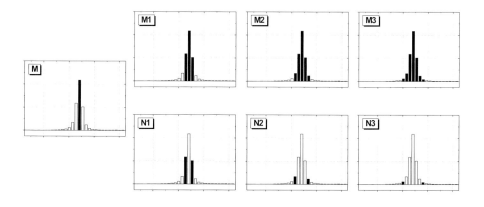

Figure 4.3: Definition of center pixel (**M**), peak volume (**M**$_x$) and side pixel (**N**$_x$) registration

65

4. *Special Features of HR-CS AAS*

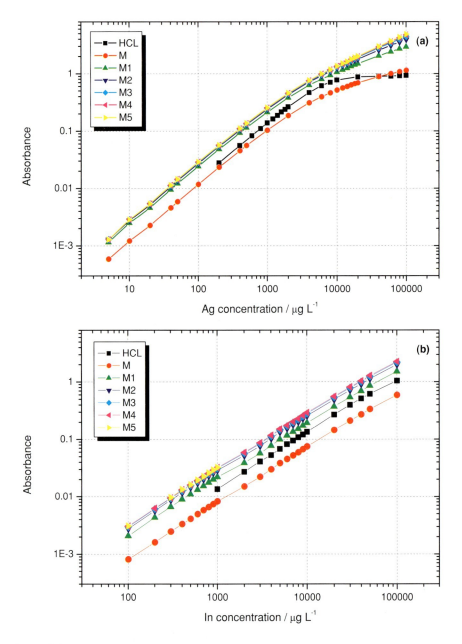

Figure 4.4: Influence of the peak volume registration scheme on the sensitivity and working range in the case of (a) silver at 328.068 nm and (b) indium at 303.936 nm

In this case a significant improvement in sensitivity can be obtained for HR-CS AAS not only by using CP ± 1, but also by further increasing the peak volume to CP ± 2 and CP ± 3. For exact comparison the LS AAS measurement was obtained with the same spectrometer setup by replacing the xenon short-arc lamp with a conventional HCL (refer to Figure 3.5).

On the other hand, using side pixels only at increasing distance to the line core for evaluation, results in reduced sensitivity and increasing linearity, as shown in the linear plots of Figure 4.5. Although the increase in the linear range, compared to LS AAS is clearly more pronounced for narrow absorption lines, mostly because of the more pronounced effect of line shifts in LS AAS, as shown in Figure 4.5 (a), the overall trend is the same in HR-CS AAS, independent of the line shape, i.e. measurement at the line wings might be used without compromise for the determination of high analyte concentrations, extending the working range of HR-CS AAS to at least 5 orders of magnitude in concentration, as claimed for ICP OES and ICP-MS.

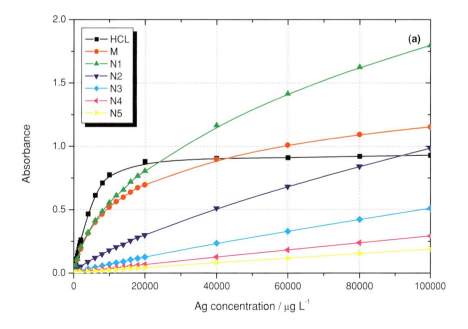

Figure 4.5: Influence of the side pixel registration scheme on the sensitivity and linearity in the case of (a) silver at 328.068 nm and (b) indium at 303.936 nm (refer to the next page)

4. Special Features of HR-CS AAS

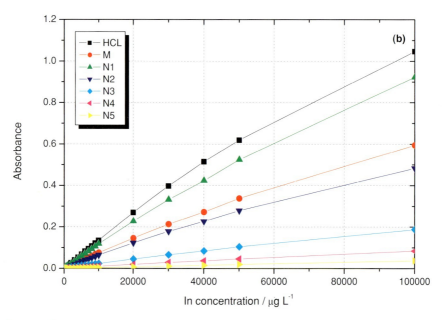

Figure 4.5: (continued); Side pixel registration scheme in the case of (b) indium at 303.936 nm

4.5 Signal-to-Noise Ratio, Precision and Limit of Detection

As the 'signal', i.e. the sensitivity has been fully treated in Section 4.4, only the 'noise' component of the above terms has to be discussed here. In addition, only the contribution of instrumental components to the noise will be treated, not the 'analytical' part, such as the imprecision of sample introduction, the flame noise, or the inhomogeneity of the sample itself, which obviously also make important contributions.

The noise contribution of the detector has already been treated in Section 2.2.2, where it has been shown that for a CCD the absorbance noise is independent of the spectral bandwidth (Equation 2.31), but it depends on the number of measurement pixels n_{sam} and reference pixels n_{ref} in such a way that n_{sam} should be as small as possible and n_{ref} should be larger than n_{sam}. The other component that influences the noise, according to Equation 2.31), is the radiance L_λ of the radiation source, in such a way that the minimal detectable volume concentration of absorbing atoms N_{min} is inversely proportional to the square root of L_λ. As the intensity of the source in CS AAS is some 1–2 orders of magnitude higher than that of typical line sources for conventional AAS, an improvement in the SNR and LOD by factors of 3–10 could be expected, unless other factors, such as the flame noise become dominant. The values for a selection of elements given in Table 4.1 show that this expectation has in fact been realized for the majority of the elements, at least in the case of the LOD [61].

Table 4.1: Overview of the characteristic concentrations and LOD (3σ) obtained with the DEMON based HR-CS AA spectrometer in an air/acetylene flame

Element	Wavelength / nm	Characteristic concentration / $\mu g\,L^{-1}$	LOD / $\mu g\,L^{-1}$
Ag	324.068	21	0.9
As	193.696	610	350
Au	242.795	60	2
Bi	223.061	110	11
Ca	422.673	33	0.7
Cd	228.802	10	0.7
Co	240.725	34	1.9
Cr	357.868	40	1.4
Cu	324.754	18	0.6
Fe	248.327	44	1.7
Ga	287.424	940	25
In	325.609	150	8
Ir	208.882	3 200	400
Li	670.785	6	0.14
Mg	285.213	3	0.09
Mn	279.482	12	1.2
Ni	232.003	40	6
Pb	217.001	73	12
Pd	247.642	100	10
Pt	265.945	690	28
Rh	343.489	160	4
Ru	349.895	800	20
Sb	217.581	200	23
Se	196.026	270	150
Sn	224.605	1 300	110
Te	214.281	180	27
Tl	276.787	125	7
Zn	213.856	7	1.1

4. Special Features of HR-CS AAS

Besides the intensity of the radiation source, its stability obviously also plays an important role, and arcs are notorious for their instability. However, as this instability results in a wavelength-independent, 'white' noise, which causes an identical noise-over-time distribution on all the pixels of the CCD array, it can easily be cancelled out with the use of reference pixels, as has already been discussed in Section 4.2. As a result of the almost perfect correlation of the spectral intensity values within the small range of observation, the minimum detectable absorbance signal is determined only by statistical variations in the intensity between the neighboring pixels (shot-noise). This means that an increase in illumination time or in radiation intensity by a factor of 4 will reduce the absorbance noise by a factor of 2 (square root of 4).

The first effect, i.e. the influence of the illumination time is shown in Figure 4.6 for the element lead at 217.001 nm. Simply by increasing the illumination time from 5 s to 45 s, i.e. by a factor of 9, the noise level (standard deviation as sum over 3 pixels) is reduced by a factor of 2.9, which is very close to the expectation for shot-noise limited measurements, and the LOD is improved by the same factor.

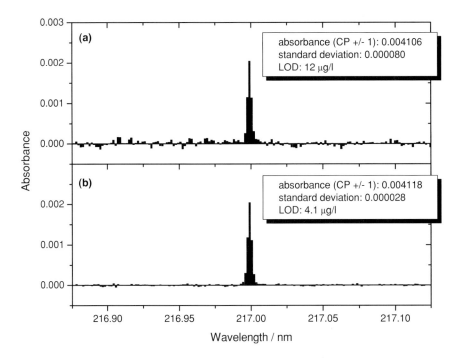

Figure 4.6: Influence of the illumination time on the absorbance noise in HR-CS F AAS for lead at 217.001 nm; (a) illumination time: 5 s; (b) illumination time: 45 s

The second effect, i.e. the influence of the radiation intensity is shown in Figure 4.7. Although the intensity distribution of the xenon short-arc lamp is almost flat over the entire spectral range, there are, in any optical system, increasing losses of radiation intensity in the far-UV range, particularly below 200 nm, that cannot be avoided [61].

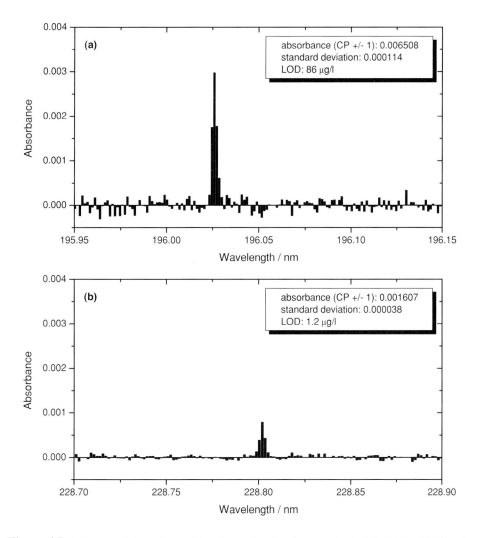

Figure 4.7: Influence of the radiation intensity on the absorbance noise in HR-CS F AAS (illumination time of 5 s); (a) determination of selenium at 196.026 nm; sample: 1000 µg/L; (b) determination of cadmium at 228.802 nm; sample: 10 µg/L

Nevertheless, as shown in Figure 4.7 (a), even for selenium at 196.026 nm, an absorbance noise (calculated in the vicinity of the absorption line) as little as 0.000114 can be obtained using an illumination time of only 5 s. For the wavelength of cadmium at 228.802 nm, which is still in the far-UV region, the noise is reduced to a value of 0.000038, i.e. a factor of three lower, due to the higher radiation intensity that reaches the detector, as shown in Figure 4.7 (b). This tendency continues towards longer wavelengths although it becomes less and less pronounced.

4.6 Multi-element Atomic Absorption Spectrometry

Harnly [50, 54] has been speculating repeatedly about simultaneous multi-element AAS, because the idea of combining the analytical capabilities of HR-CS AAS with simultaneous multi-element detection is intriguing. In essence, any instrument that uses a high-resolution echelle spectrometer and a sufficiently intense continuum source, such as a xenon short-arc lamp, is capable of simultaneous multi-element AAS measurement. A spectrometer of this type, the ARES (see Section 3.2.3), has actually been built by Becker-Ross and co-workers [9] as a tool for studies of structured background in F AAS, although it did not cover the entire wavelength range of AAS. The major limitation of this instrument is the read-out time of the detector of 2 s per spectrum; in order to obtain a good SNR the authors used an alternate measurement of 50 spectra of both sample and blank solution, resulting in a total measurement time of several minutes. This might be acceptable for F AAS, although it would not bring any advantages in speed of analysis compared to conventional LS AAS using fast sequential techniques, and it cannot be used at all for transient signals, as they are produced in GF AAS. The aspect of simultaneous multi-element AAS, and its potential future development will therefore be discussed in Chapter 9.

However, any spectrometer that uses a continuum source and a double monochromator with an echelle grating makes it possible to reach any line within an extremely short period of time, much less than 1 s, as both the grating and the prism are stepper-motor controlled. This feature allows one to perform a fast sequential multi-element determination with the great advantage that flame conditions and burner height can also be optimized under computer control for each element. This means that each element can be determined under optimized not compromised conditions, as would be the case for simultaneous multi-element determination. The high intensity of the xenon short-arc lamp, which results in high SNR, further supports fast sequential multi-element F AAS measurements, as it makes possible very short illumination times, so that stabilization of the flame becomes the rate-determining factor in cases where a change of flame stoichiometry or of flame gases (air to nitrous oxide) is necessary. Nevertheless, the total time period required for changing the wavelength, adjusting flame stoichiometry and burner height, and for measuring the absorbance signal, should typically be of the order of a few seconds only. Obviously, the initial calibration of the equipment, and an eventual re-calibration after a set of measurements, will require slightly more time than the following determination.

The typical routine determination of a number of elements in a set of similar sample solutions will therefore no longer be the determination of element A in all the samples, followed by a change of radiation source, wavelength, flame conditions, burner height etc., and determination of element B in the same samples, and so on, as is common practice in LS F AAS. In HR-CS F AAS it will rather consist of a calibration of the instrument for all elements of interest, followed by a determination of all elements in sample 1, all elements in sample 2, all elements in sample 3, etc. It may be worth mentioning that, at least for a limited number of 10–15 elements, the total analysis time required for HR-CS F AAS will be shorter than that for even a simultaneous ICP OES measurement because of the much shorter equilibration time required for a typical AAS burner after changing the sample solution, compared to that for the spray chambers used in ICP OES.

Another option that is available with HR-CS AAS to a much greater extent than with LS AAS is the use of secondary, less sensitive analytical lines, as already discussed in Section 4.2, because all lines have essentially the same high SNR, i.e., there are no more 'weak lines', although the emission intensity and the detector sensitivity decrease at wavelengths below 220 nm. This means that two or three analytical lines of the same element with significantly different sensitivity can be chosen in the fast sequential mode in cases where the analyte concentration in the samples is unknown, and might change over several orders of magnitude. In this way the analyte can always be determined in the optimum working range of an analytical line without the need for any additional dilution of the measurement solution. This obviously results in a significant time saving, and avoids the risk of dilution errors.

Yet another feature that becomes available with the fast sequential mode of operation in HR-CS AAS is the use of the 'reference element technique', i.e. the use of an 'internal standard'. This technique has been described extremely rarely in LS AAS, although several papers appeared, mainly in the 1970s, which are summarized in a review article by Dulude and Sotera [27]. The reasons are obvious: Firstly, the technique requires a dual- or multi-channel spectrometer, and there have been only very few spectrometers of this type available commercially over the past decades. Secondly, the reference element technique is ideally suited to correction for non-specific interferences, such as transport interferences, but it is notoriously difficult to find an appropriate reference element for element-specific interferences. Thirdly, the most successful multi-channel LS AAS equipment, the Perkin-Elmer model SIMAA 6000, which has been available for a number of years, was designed for GF AAS only, a technique that does not typically exhibit non-specific interferences. The number of publications using this technique is therefore very limited [17, 33, 112, 118].

In HR-CS AAS, although no simultaneous measurement of two elements is yet possible, the reference element technique can be used essentially without compromises for F AAS, where non-specific interferences are quite common. The fast-sequential mode of operation can handle this type of interference without problems as it does not change with time. The most widespread interference of this type is transport interference caused by changes in the viscosity between sample and calibration solutions and from sample to sample due to variations in the matrix composition. In LS FAAS these interferences are usually eliminated by 'matrix matching' if the composition of the sample matrix is known and constant for a large number of samples, or by the analyte addition technique when the sample matrix is unknown or differs from sample to sample. The former technique usually requires a high concentration of ultra-pure chemicals in order to keep the risk of introducing high blank values as low as possible. The latter technique is very labor-intensive and time-consuming, it is strictly limited to the linear part of the calibration curve, and is characterized by inferior precision, because the analyte content is determined by extrapolation, not interpolation. All these limitations are absent when the reference element technique is used to eliminate non-specific interferences in F AAS, whereby the timesaving aspect is probably the most significant.

Another advantage of the reference element technique is that dilution errors can be recognized and eliminated when the reference element is added at an early stage of sample preparation before the final dilution. Moreover, using the reference element technique, quantitative analysis could actually be done without accurate dilution and without using volumetric flasks etc., resulting in additional timesaving and elimination of potential errors in sample preparation.

4.7 Absolute Analysis

Although AAS, like all other spectrometric procedures, is in principle a relative technique that can only provide quantitative results through the use of calibration samples, the possibility of 'absolute analysis', i.e. the calculation of the analyte quantity directly from the absorbance using physical constants, has been discussed again and again. Even in his first paper in 1955, Walsh [146] pointed out the theoretical possibility of performing absolute analysis by AAS. The use of a flame as the atomizer, however introduced too many variables to come even close to an absolute analysis. Some two decades later, L'vov [95] established the basis of modern GF AAS and proposed in the title *'Electrothermal atomization – the way towards absolute methods of atomic absorption analysis'* that absolute analysis has come within reach using this technique. Slavin and Carnrick [134], a few years later, were able to show that using the STPF concept, i.e. L'vov's ideas, GF AAS offered the possibility of performing determinations with an accuracy of $10-20\%$ in varying and complex matrices without calibration, using the characteristic mass m_0 as the reference quantity. L'vov et al. [97] then attempted to calculate m_0 on the basis of a simplified model. The major sources of variations in sensitivity were due to the age and individual characteristics of line sources and operating conditions.

For this reason, Gilmutdinov and Harnly [42] brought up the idea again, proposing that HR-CS AAS is more suitable for absolute analysis than LS AAS. They showed that integration in absorbance is a prerequisite for absolute measurements, and that absorbance has to be resolved in wavelength, space and time. Such measurements can obviously be made simultaneously with a continuum source, a high-resolution echelle spectrometer and a solid-state array detector. It is obvious that this approach is a further step towards the realization of absolute analysis, as it eliminates some of the most notorious variables in the concept, although it has been criticized by L'vov [99].

Due to its notable stability even in LS AAS, the characteristic mass m_0 is an extremely useful control criterion in GF AAS that is universally recognized. Within a laboratory the m_0 can serve to check the correct functioning of an instrument and the correctness of the selected parameters. It can also provide information on trueness during the development of new methods and serve as a criterion in interlaboratory trials. It might be expected that the stability of the characteristic mass m_0 will further increase in HR-CS AAS. Independent of the recognized usefulness of m_0, the topic of 'absolute analysis' will remain a topic of controversial discussion, and be more of academic than of practical interest. The latter is mainly due to the fact that no laboratory can really do without calibration samples and calibration solutions, at least to check m_0. Quite apart from the fact that nowadays the majority of laboratories are required by regulation to use a given number of calibration samples at given intervals to verify the accuracy of their analytical results.

5. Measurement Principle in HR-CS AAS

5.1 General Considerations

In practical analytical work radiation is not only absorbed by atoms, but also by gaseous molecules and radicals, and radiation may also be scattered at particles in the atomizer. As outlined in Section 2.3, molecular absorption spectra in the visible and UV range are due either to electron excitation and exhibit a pronounced rotational fine structure, or to molecule dissociation and are then 'continuous', i.e. without fine structure, and they do not change, or change monotonously, within the observed spectral interval. Molecular absorption may be caused by concomitants in the sample or by the atomizer itself, e.g. the flame gases. Radiation scattering may be observed mainly in graphite tube atomizers when refractory matrix components cannot be removed during the pyrolysis stage. Radiation scattering obviously is also 'continuous', i.e. without fine structure. All this attenuation of radiation, which is not specific to the analyte, is usually termed 'background' or 'non-specific' absorption.

Measurement of and correction for background has always been one of the major challenges in all spectrometric techniques, as is illustrated in the drawing of (the late) Hans Massmann, which is shown in Figure 5.1. As it is impossible to measure atomic absorption only, it is necessary to measure total absorption first, then background absorption, and to subtract the latter from the former. In conventional LS AAS these two measurements have always to be made sequentially, and in the case of dynamic signals, as in GF AAS, measurement frequency plays a crucial role. Holcombe and Harnly [65] have treated the measurement errors and artifacts that might be encountered with rapidly changing background signals.

There are several techniques available with LS AAS to measure and correct background absorption, such as deuterium (D_2) background correction (BC), Zeeman-effect

5. Measurement Principle in HR-CS AAS

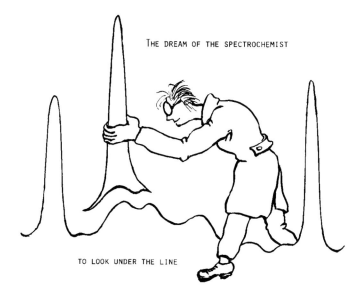

Figure 5.1: 'The dream of the spectrochemist – to look under the line'
Drawing of (the late) Hans Massmann, ISAS – Institute for Analytical Sciences, Germany (former: Institute of Spectrochemistry and Applied Spectroscopy)

BC, and BC using high-current pulsing. These BC techniques are fully treated in the book of Welz and Sperling [150]. All these techniques have their limitations, and all can actually introduce errors when they are applied without knowing exactly the nature of the background absorption – which is almost impossible to discover with LS AAS. All techniques can handle 'continuous' background reasonably well, unless it is changing too rapidly (see above), and as long as it is not too high. The limits for deuterium BC are reached at a background absorbance between 0.5 and 1, whereas the other techniques can correct for background absorbance up to about 2.

All these techniques, however, have problems with structured background, i.e. with electron excitation spectra. Deuterium BC measures and subtracts a background 'averaged' over the spectral bandwidth given by the monochromator, which almost never corresponds to the actual background at the analytical line, resulting in over- or under-correction, i.e. in a measurement error. BC with high-current pulsing measures the background on both sides of the analytical line, which in the case of structured background cannot be the same as at the analytical line, resulting in the same problem. Zeeman-effect BC, the most powerful of the three techniques, can actually correct for fine-structured background, as it measures total and background absorption in the same spectral interval with the same line profile. A precondition for an exact correction is however that the background absorption is not affected by the magnetic field, which unfortunately is not necessarily the case, as has been shown by several authors, particularly for the PO molecule [15, 57, 58, 111, 156, 160]. Another source of error using Zeeman-effect BC is when another atomic line is in the vicinity (within about ± 10 pm) of the analytical line, as the σ-components of this line might overlap with the analytical line and cause interference [15, 39, 152–155].

5.2 Background Measurement and Correction

There are at least two distinct differences in the measurement of background in HR-CS AAS compared to LS AAS. Firstly, due to the wavelength-resolved detection of the absorbance, the background, its nature and spectral distribution become visible, which makes it orders of magnitude easier to find the appropriate action for its removal or correction. Secondly, measurement of atomic and background absorption is strictly simultaneous, so that no artifacts or 'bracketing' effects, such as those described by Holcombe and Harnly [65] are observed, even for the most rapidly changing background signals.

5.2.1 Continuous Background

The computer-aided signal processing of an analytical cycle always has the following simplified scheme: Initially, a dark spectrum is recorded by blocking the beam path of the radiation coming from the xenon short-arc lamp with a mechanical shutter. Then, either before starting the temperature program in the case of a graphite furnace measurement, or during the aspiration of the blank in the case of a flame measurement, a given number n_{BOC} of successive reference intensity spectra I_{BOC}^{pixel} without any analyte absorption, the so-called background offset correction (BOC) scans, is recorded. These scans are normalized according to their sum intensity over all pixels, subsequently averaged, and used as the averaged reference intensity values $\bar{I}^{pixel}$ of the individual pixels:

$$\bar{I}^{pixel} = \frac{1}{n_{BOC}} \sum_{BOC} I_{BOC}^{pixel}. \tag{5.1}$$

This spectrum represents the actual intensity distribution at the specific wavelength position, which is mainly given by the intensity spectrum of the continuum source, the blaze characteristic of the echelle grating, and the individual sensitivity of the CCD pixels. A model spectrum showing a typical averaged intensity spectrum, including a pixel error on the left-hand side, is depicted in Figure 5.2.

5. *Measurement Principle in HR-CS AAS*

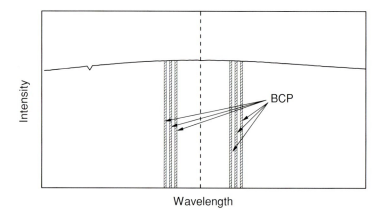

Figure 5.2: Model of an averaged intensity spectrum, including a pixel error and a selection of background correction pixels (BCP); analyte wavelength position: dashed line

Next, a certain number of analyte intensity scans is recorded, giving the three-dimensional plot, intensity versus wavelength and time, shown in Figure 5.3. In total, three different types of absorption signals can be observed: continuous background absorption, analyte absorption, and fine-structured atomic or molecular absorption from concomitants.

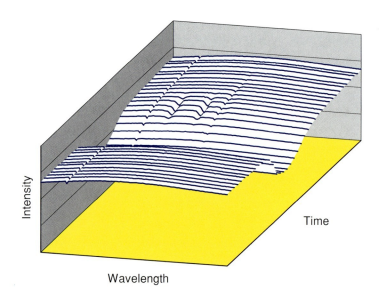

Figure 5.3: Model of a series of uncorrected analyte intensity spectra

5.2 Background Measurement and Correction

There are two possibilities to eliminate the continuous background absorption part. The first is based on the selection of so-called background correction pixels (BCP). The intensities of these particular pixels are summed and used for the correction by calculating the ratio of the sum intensities ($\overline{I}_{BCP}$ and $I_{BCP,scan}$) of the BCP in the averaged reference spectrum $\overline{I}^{pixel}$ (refer to Figure 5.2) and in each individual analyte intensity spectrum I_{scan}^{pixel}, respectively. This gives a correction factor α_{scan} for the individual scan:

$$\alpha_{scan} = \frac{\overline{I}_{BCP}}{I_{BCP,scan}} = \frac{\sum_{BCP} \overline{I}^{pixel}}{\sum_{BCP} I_{scan}^{pixel}} . \tag{5.2}$$

This intensity correction factor is a measure of the continuous background absorption, leading to a decrease in the total signal intensity within the recorded spectral interval (see Figure 5.3) and is quite similar to the non-specific background value provided by conventional LS AAS systems, as shown in Figure 5.4.

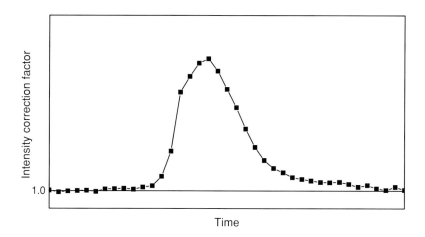

Figure 5.4: Temporal behavior of the intensity correction factor as a measure of the continuous background absorption

Using the correction factor, each analyte intensity spectrum can be normalized to the averaged reference spectrum by:

$$I_{norm,scan}^{pixel} = \alpha_{scan} \, I_{scan}^{pixel} . \tag{5.3}$$

Taking the ratio between the averaged reference spectrum and each normalized analyte intensity spectrum gives the individual transmittance spectrum. The total series is depicted in Figure 5.5.

5. Measurement Principle in HR-CS AAS

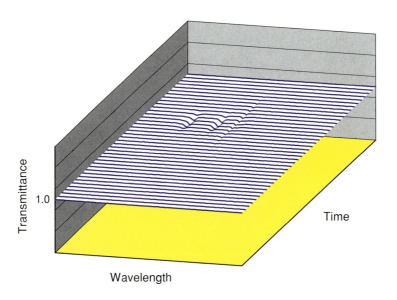

Figure 5.5: Series of continuous background corrected transmittance spectra, calculated from Figure 5.3

Finally, the calculation of the logarithm of the ratios leads to the individual absorbance spectra $A_{\text{scan}}^{\text{pixel}}$:

$$A_{\text{scan}}^{\text{pixel}} = \log \frac{\overline{I}^{\text{pixel}}}{I_{\text{norm,scan}}^{\text{pixel}}} \ . \quad (5.4)$$

In this way, continuous background absorption is removed and, in addition, all systematic errors, such as pixel errors and intensity fluctuations of the continuum source, are also eliminated, resulting in the three-dimensional plot of absorbance versus wavelength and time, shown in Figure 5.6, where only atomic and fine-structured absorption remains.

To force the absorbance spectra to the zero line requires the correct choice of the BCP, which is, of course, a critical process since the BCP must not be influenced by other absorption effects. As a measure of their suitability as correction pixels one can, for instance, use the logarithm of the ratio of the intensities at two neighboring pixels not too close together [52]:

$$\gamma = \log \frac{I_{\text{scan}}^{\text{pixel}}(\text{pixel 1})}{I_{\text{scan}}^{\text{pixel}}(\text{pixel 2})} \ . \quad (5.5)$$

In the ideal case where the behavior of the two pixels is synchronous γ is constant. Large deviations from a constant value, on the other hand, indicate that at least one of the pixels is unsuitable as a BCP.

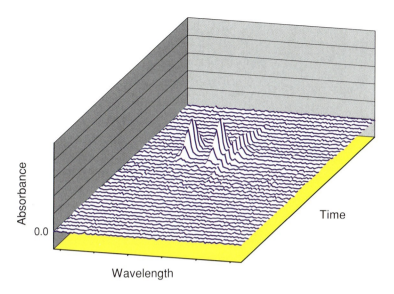

Figure 5.6: Series of absorbance spectra corrected for continuous background and other continuous events, calculated from Figure 5.3

While it is possible using the BCP method to compensate for background that is constant over the recorded spectral interval, 'sloping' background, i.e. broadband absorption that changes in wavelength within the observed interval, cannot be entirely corrected. In this case best correction using BCP is obtained when the correction pixels are chosen to be symmetrical with regard to the analyte wavelength. A practical example of such a situation will be discussed in Section 8.2.9.

The second method for the correction of continuous, and especially of sloping background, is to bring the absorbance spectra to the zero line using the linear least-squares fitting of a polynomial. This procedure is no longer done in intensity, but in the absorbance space. Suitable algorithms may be found in publications such as [117].

By fitting the minima of each absorbance spectrum to a polynomial one gets complete independence from the BCP. However, in spite of getting rid of the BCP selection, the critical point now is the correct choice of the minima. This is accomplished by dividing the recorded spectral interval into a certain number of windows. From each window the local minima are selected as sampling points for the polynomial fit. Using only the absolute minima for the calculation would result in over-correction. As a fluctuation of the baseline around the zero line would be correct, not only the absolute minimum but also all values within a certain noise band above the minimum of each window have to be included. The noise band is obtained from five times the standard deviation σ_{BOC} which is calculated from the BOC scans averaged over all pixels.

In addition, it has to be considered that the intensity is in general highest during the BOC scans, and the shot-noise therefore is at its minimum. Since the noise level increases with the decrease in intensity resulting from continuous background, the width of the noise band for each scan σ_{scan} has to be corrected for, according to the intensity correction factor α_{scan} of the individual scan:

$$\sigma_{scan} = 5\,\sigma_{BOC}\,\sqrt{\alpha_{scan}}\,. \tag{5.6}$$

Taking this into account, all pixels within a window with a value for uncorrected absorbance inside the noise band are included in the calculation. From these values the mean value of both absorbance values and positions is obtained; thus for each window a mean absorbance value is correlated with a mean wavelength position. In this way, a certain number of sampling points is obtained for the polynomial fit (refer to Figure 8.52).

In LS GF AAS, according to the *Stabilized Temperature Platform Furnace* (STPF) concept [133], signal evaluation should be done exclusively by means of time-integrated absorbance, the so-called peak area value. Applying this principle to HR-CS GF AAS, a spectrum is obtained which is summed over time, whereby the integrated absorbance at individual pixels A_{int}^{pixel} is calculated by:

$$A_{int}^{pixel} = \frac{1}{\Delta t_{illu}} \sum_{scans} A_{scan}^{pixel}, \tag{5.7}$$

where Δt_{illu} is the illumination time per scan of the CCD detector. The resulting time-integrated absorbance spectrum is shown in Figure 5.7.

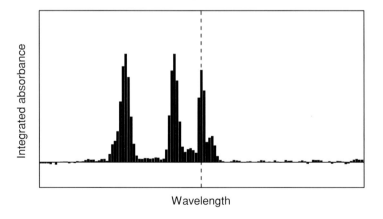

Figure 5.7: Time-integrated absorbance spectrum, calculated from Figure 5.6; analyte wavelength position: dashed line

This kind of spectrum is comparable to that recorded in HR-CS F AAS. There, of course, the absorbance values are averaged in time over all scans for the individual pixels, resulting in an absorbance-over-wavelength plot. Having only one absorbance spectrum the time-consumption for signal processing as well as the complexity of the fitting algorithms are reduced dramatically, since it is not necessary to apply all the fitting procedures to the individual scans, as in the case of transient signals.

Another fact should be mentioned here: Because of the high spectral resolution provided by the spectrometer described in Section 3.2.2, in most cases signal detection should not necessarily be done only on a single pixel, the center pixel. As shown in Figure 5.7, the analyte absorbance line stretches over at least three pixels. Therefore, the inclusion of part of the line wings, the so-called peak volume registration, should be taken into account for maximum sensitivity. For more details about the different signal registration schemes and their advantages refer to Section 4.4.

5.2.2 Fine-structured Background

Once the continuous background is eliminated as described in the previous section, apart from the analyte absorption, the fine-structured molecular absorption and/or the eventual atomic absorption due to concomitant elements becomes visible. In most cases it is not necessary to take further action in order to remove this residual fine-structured background, as long as it does not directly overlap the analyte absorption in wavelength and time. Again, the decision can usually be made easily by inspecting the time- and wavelength-resolved absorbance readings. Obviously, an interference occurs only when the background absorption is on the analytical pixel(s), and if it is not separated in time from the analyte absorption (in the case of transient signals). The first case is relatively rare due to the high spectral resolution of the spectrometer, and the second case can be influenced by optimization of the graphite furnace program (refer to Section 8.2.1).

For the case where analyte and background absorption coincide both spectrally and temporally, as shown in the series of model spectra in Figure 5.6, it is possible to correct for fine-structured background absorption using reference spectra of the interfering molecule(s). Spectral interferences caused by other atoms because of direct line overlap will be treated in the next section.

A fundamental requirement for the correction of fine-structured background via reference spectra is that the spectrometer is equipped with an accurate mechanism for wavelength stabilization. This feature is included in all research spectrometers and described in detail in Section 3.2. At the beginning of each analytical cycle the actual wavelength position is checked and adjusted, if necessary, to ensure an exact pixel-to-wavelength correlation.

5. Measurement Principle in HR-CS AAS

In addition, the reference absorbance spectra ($A_{\text{ref},i}^{\text{pixel}}$) of the number i of molecules causing fine-structured background at the analyte wavelength position must be known. These reference spectra are then used as independent linear functions in a least-squares fitting algorithm and fitted to each absorbance spectrum. The individual absorbance spectra corrected for fine-structured background are calculated as follows:

$$A_{\text{corr,scan}}^{\text{pixel}} = A_{\text{scan}}^{\text{pixel}} - \sum_i a_{\text{scan},i} A_{\text{ref},i}^{\text{pixel}} . \tag{5.8}$$

The molecule correction factors $a_{\text{scan},i}$, which are a measure of the strength of the different reference absorbance spectra for the corresponding scan, are obtained by omission of the pixel at the analyte wavelength position and its neighbors, since the actual analyte absorption may not influence the background correction procedure.

In the case of the model measurement in Figure 5.6 the fine-structured background can be eliminated by using only one molecule reference spectrum, which is depicted in Figure 5.8. The temporal behavior of the corresponding molecule correction factor, which is provided by the least-squares fitting procedure, is shown in Figure 5.9. In comparison to the intensity correction factor used for correction of the continuous background (refer to Figure 5.4) one can see the totally different temporal behavior. In the above-described way, all background structures are successfully removed and finally the desired undisturbed absorbance spectra of the analyte are obtained, as shown in Figures 5.10 and 5.11. A practical example of correction for fine-structured background, using two molecule reference spectra simultaneously, will be discussed in Section 8.2.3.

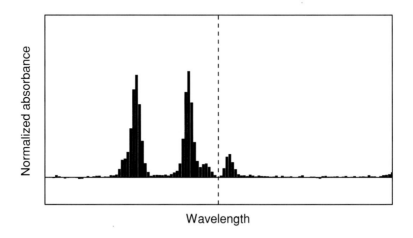

Figure 5.8: Molecule reference spectrum used for fine-structured background correction of the model measurement in Figure 5.6; analyte wavelength position: dashed line

5.2 Background Measurement and Correction

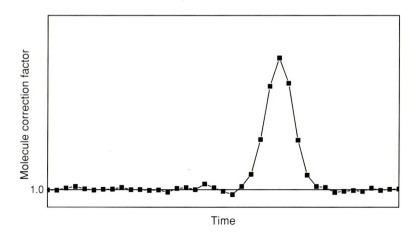

Figure 5.9: Temporal behavior of the molecule correction factor as a measure of the fine-structured background absorption caused by the molecular species of Figure 5.8 present during atomization

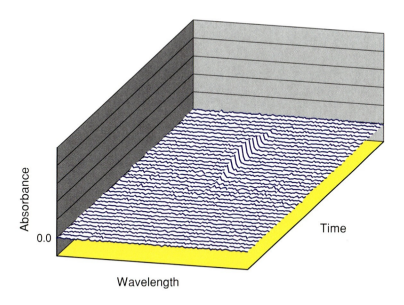

Figure 5.10: Series of absorbance spectra corrected for all background effects, calculated from Figure 5.6

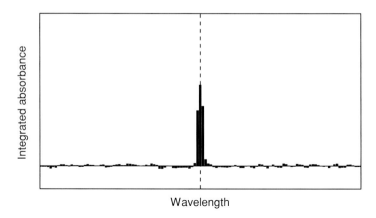

Figure 5.11: Time-integrated absorbance spectrum corrected for all background effects, calculated from Figure 5.10; analyte wavelength position: dashed line

Preferably, the molecule reference spectra, at least those of the most notorious troublemakers, such as PO and NO (refer to Chapter 7), should be pre-recorded with a good SNR and stored in a library. Fortunately, identification of the molecule responsible for the specific fine-structured background is not in all cases necessary, since a reference spectrum can be generated from the measurement itself, when the molecular structure can be separated from the analyte signal in time. By limiting the number of scans to the time interval of the presence of the unknown molecular structure, a reference spectrum can be obtained when calculating the time-integrated absorbance from the selected scans.

A special situation exists in the case of flame measurements. Here the flame gases themselves produce fine-structured background, mainly OH bands (refer to Section 7.2.13). This kind of background must be treated in a different way, since it is permanently present when recording both, the analyte intensity spectra and the BOC scans. In general permanent structures are of minor importance, since they are typically of the same order over time and will therefore be eliminated during signal processing. Nevertheless, when working close to the LOD, where the flame instabilities become more significant, or when the matrix is not matched for both recordings, the strength of the permanent background can be different and will then result in additional positive or negative structures, depending on their ratio. Fortunately, there is an elegant possibility to correct for them right at the beginning and automatically, i.e. without the need for a previously recorded molecule reference spectrum. By adding an offset to the averaged reference intensity spectrum and subsequently calculating the corresponding absorbance values from the ratio between the original and modified values, one gets a molecule reference spectrum representing the permanent structures that can be used for the least-squares fitting procedure.

5.2.3 Direct Line Overlap

The presence of a concomitant element can lead to additional absorption lines within the recorded spectral interval. Again, they are only of interest when they directly overlap with the analyte line and cannot be separated in time in the case of GF measurements.

Even in this rare case HR-CS AAS can eliminate the interference due to direct line overlap presupposing that the interfering element has at least one additional absorption line within the recorded spectral interval. Since the line strengths are directly correlated to each other, the unknown contribution of the overlapping line to the total absorbance signal at the analyte wavelength position can be determined by using the additional absorption line of the interfering element for background correction. This is accomplished in a way similar to the correction for fine-structured background resulting from electron excitation of molecules, as discussed in the previous section. After recording the absorbance pattern of the interfering element alone, this reference spectrum can be stored and used as a linear function in the least-squares fitting procedure. Finally, the undisturbed absorbance signal of the analyte can be obtained by subtracting the fitted reference spectrum from the actual absorption pattern.

An example of a direct line overlap and its elimination is given in Section 8.1.4, where the zinc concentration is determined on its resonance line at 213.856 nm in the presence of high amounts of iron, having a weak close-by absorption line at 213.859 nm.

6. The Individual Elements

Following the general strategy of this book as a kind of supplement to the book *Atomic Absorption Spectrometry* by Welz and Sperling [150], this chapter repeats as little as possible of what is already written in that book. This includes particularly information about the occurrence, use, biological significance, environmental relevance and toxicity of the individual elements, as well as of particular species of these elements. There is also no information in this chapter about optimum pyrolysis and atomization temperatures or modifiers for the individual elements when GF AAS is used for their determination. For these details as well as for information about the stability of solutions, the volatility of individual compounds or the risk of contamination the authors ask the reader to refer to Reference [150].

This chapter concentrates almost exclusively on information related to HR-CS AAS, such as the characteristic concentration c_0, i.e. the analyte concentration that results in an absorbance signal of 0.0044 with flame atomization, the characteristic mass m_0, i.e. the analyte mass that results in an integrated absorbance signal of 0.0044 s for electrothermal atomization, and the linear working range for the most sensitive analytical line, in cases where it is significantly different from LS AAS, using CP $\pm$ 1 for all measurements.

The data for about 50 elements were collected; for the rest of the analytes, mostly rare-earth elements, please refer to Reference [150]. For almost 30 of these elements, additional information has been compiled in extensive tables, containing secondary lines and their relative sensitivities in comparison to the main analytical line, as these lines can be used without compromises in HR-CS AAS. The selection of secondary lines was made with the goal of covering a sensitivity range of about two orders of magnitude. In principle, this range could be extended to more than three orders of magnitude for many elements. The example in Figure 6.1 shows several Rydberg series for aluminum, converging to the ionization limit, with the weakest lines at 207.71 nm showing sensitivities more than a factor of 6000 lower, than the main analytical line.

The major and secondary atomic lines are listed together with those of other elements which are within $\pm\lambda/1000$ (i.e. ± 140 pixels of the applied CCD detector) of the respective analytical line, and which might hence appear within the observed spectral range.

6. The Individual Elements

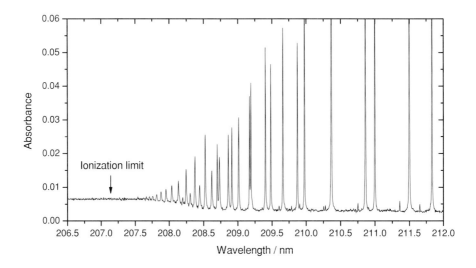

Figure 6.1: Enlarged spectrum of the Rydberg levels and ionization limit of aluminum in the range 206.5 nm to 212.0 nm measured in absorption with a nitrous oxide / acetylene flame (sample: 5 g/L Al; reference: deionized H_2O)

This might be used as information about major matrix elements present in the sample of investigation. More importantly, however, these lines might be used for the simultaneous determination of additional elements together with the main analyte, if the sensitivity ratios are favorable. For this reason the relative sensitivity for all these lines in comparison to the main analytical line of the respective elements is given in order to allow a preliminary estimation of whether a simultaneous determination is feasible within the concentration range of interest.

Absorption lines of matrix elements that are within ±10 pixels of the respective analytical line are printed in **bold** in order to warn of the potential occurrence of a spectral interference. A general warning is given in the text if the main analytical line is in the vicinity of the most common molecular absorption bands, i.e. of OH, CN, NO or PO. Information about single absorption 'lines' of these molecules is given in the table when the absorption 'line' is within ±10 pixels of the respective analytical line, and hence requires special attention. These 'lines' are printed in ***bold italic***.

For more information about molecular structures that might be observed in flames under given conditions please refer to Chapter 7. Interferences that might be observed in the presence of specific matrices, and their elimination in HR-CS AAS, are treated in Chapter 8.

A summary of the main parameters, e.g. the characteristic concentrations (c_0) with flame atomization, characteristic masses (m_0) with electrothermal atomization and maximum absorbance (A_{max}) for linearity, at the most sensitive absorption line of the individual elements, which are discussed in detail in the following sections, is given in Table 6.1.

Table 6.1: Comparison of the characteristic concentrations (c_0) with flame atomization, characteristic masses (m_0) with electrothermal atomization and maximum absorbance (A_{max}) for linearity at the most sensitive analytical line of the elements using HR-CS AAS

Element	Wavelength / nm	Characteristic concentration / $\mu g\, L^{-1}$	Characteristic mass pg	A_{max} / mg L^{-1}
Ag	328.068	21	3	
Al	309.271	370	15	
As	193.696	610	50	
Au	242.795	60	11	3.3
B	249.772	4 400	76	
Ba	553.548	140	7	
Be	234.861	6	1.3	
Bi	223.061	110	16	7
Ca	422.673	33	1.2	
Cd	228.802	10	0.8	0.4
Co	240.725	34	7	1.3
Cr	357.868	40	4	1.0
Cs	852.118	65	10	18
Cu	324.754	18	6	2.4
Eu	459.403	150	9	
Fe	248.327	44	8	1.8
Ga	287.424	940	20	80
Ge	265.157	850	55	
Hg	253.652		160	
In	303.936	150	12	33
Ir	208.882	3 200	120	55
K	766.491	5	1.2	
La	550.134	45 000	7 000	
Li	670.785	6	1.0	
Mg	285.213	3	0.5	0.25

6. The Individual Elements

Table 6.1 (continued)

Element	Wavelength / nm	Characteristic concentration / $\mu g\,L^{-1}$	Characteristic mass pg	A_{max} / $mg\,L^{-1}$
Mn	279.482	12	1.8	
Mo	313.259	390	13	
Na	588.995	2.4	0.2	0.35
Ni	232.003	40	10	
P (via PO)	324.619	70		
Pb	217.001	73	6	
Pd	247.642	100	22	
Pt	265.945	690	110	
Rb	780.027	14	1.6	3.3
Rh	343.489	160	23	
Ru	349.895	800	40	
S (via CS)	257.594	90 000		3 000
Sb	217.581	200	21	
Se	196.027	270	40	
Si	251.611	1 200	70	
Sn	224.605	1 300	40	
Sr	460.733	80	1.2	
Te	214.281	180	17	
Ti	319.200	1 000	45	
Tl	276.787	125	10	
V	318.397	280	20	
W	255.135	6 000		
Zn	213.856	7	0.5	

6.1 Aluminum (Al)

Aluminum is the third most abundant element in the Earth's crust, and its omnipresence, e.g. in dust, results in an extremely high risk of contamination, particularly when traces of this element have to be determined. The most sensitive analytical line for aluminum is at 309.271 nm with a characteristic concentration of $c_0 = 0.37$ mg/L in a nitrous oxide / acetylene flame, which is about a factor of two better than what is obtained in LS AAS. The characteristic mass at this line, using a transversely heated graphite tube atomizer, is

$m_0 = 15$ pg, which is almost four times better than that obtained in LS AAS. This line is within the range of strong OH absorption bands, which might cause some deterioration of the SNR if not corrected for.

Figure 6.2 shows that a second aluminum line at 309.284 nm is located very close to the main line at 309.271 nm. Although the two lines are not completely resolved, the evaluation of the strong line is not markedly disturbed in HR-CS AAS, whereas the presence of this second line contributes to the non-linearity of the calibration curve in LS AAS.

A number of less sensitive analytical lines that might be used for the determination of higher aluminum concentrations are listed in Table 6.2, together with information about their spectral environment.

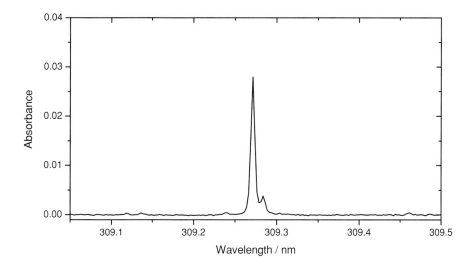

Figure 6.2: Primary absorption line of aluminum at 309.271 nm and the secondary line at 309.284 nm

Table 6.2: Selected absorption lines of aluminum (sorted according to their relative sensitivity) and corresponding spectral environment in order to identify species with potential risk for spectral interference, or additional elements that might be determined simultaneously

$\lambda_{element}$ / nm	Rel. sens.	Matrix name	λ / nm	Rel. sens.	$\lambda_{element}$ / nm	Rel. sens.	Matrix name	λ / nm	Rel. sens.
309.271	1.0	La	308.984	9.9			**OH**	**309.279**	
		V	309.144	350			V II	309.311	

Table 6.2 (continued)

$\lambda_{element}$ / nm	Rel. sens.	Matrix name	λ / nm	Rel. sens.	$\lambda_{element}$ / nm	Rel. sens.	Matrix name	λ / nm	Rel. sens.
		W	309.350	55			*NO*	**236.709**	
396.152	1.1	Ti	395.821	3.2			W	236.769	18
		Ti	396.285	18			Ti	236.856	400
		Ti	396.427	16			Ti	236.928	530
308.215	1.6	Ni	308.075	200	226.346	5.7	Ni	226.143	110
		OH	**308.206**				Ti	226.178	610
		V	**308.211**	**33**			Co	226.260	680
		Co	308.261	110			*NO*	**226.343**	
		W	308.491	140			**Fe**	**226.347**	**5000**
394.401	2.2	Ti	394.777	15			Al	226.374	140
237.312	2.3	W	237.088	35			W	226.388	12
		Ga	237.129	100			Ti	226.402	17
		Fe	237.143	660			V	226.440	380
		Co	237.185	230			Fe	226.505	980
		Ti	237.196	80	257.509	5.7	V	257.402	32
		Al	237.207	120			Co	257.435	64
		Ti	237.225	670			Al	257.540	53
		Co	237.283	510			Mn II	257.610	
		PO	**237.299**				V	257.729	250
		PO	**237.321**		221.006	5.8	Co	220.786	260
		Al	237.335	21			Si	220.798	14
		Fe	237.362	120			Mn	220.881	75
		W	237.447	9.3			W	220.906	180
		Ti	237.459	180			Sn	220.965	7.6
		W	237.476	34			*NO*	**221.005**	
236.705	4.1	Cr	236.473	82			Tl	221.065	140
		Co	236.506	51			Tl	221.079	300
		W	236.545	15			Si	221.089	5.9
		Cr	236.591	120			Ni	221.101	760
		W	236.618	66			In	221.114	620
		Cr	236.681	150			Ni	221.129	700
		PO	**236.686**				Si	221.174	17
		PO	**236.709**				Ni	221.216	920

Table 6.2 (continued)

$\lambda_{element}$ / nm	Rel. sens.	Matrix name	λ / nm	Rel. sens.	$\lambda_{element}$ / nm	Rel. sens.	Matrix name	λ / nm	Rel. sens.
237.335	21	Ga	237.129	100			Mn II	257.610	
		Fe	237.143	660			V	257.729	250
		Co	237.185	230			W	257.770	220
		Ti	237.196	80	237.207	120	Co	237.051	210
		Al	237.207	120			W	237.088	35
		Ti	237.225	670			Ga	237.129	100
		Co	237.283	510			Fe	237.143	660
		Al	237.312	2.3			Co	237.185	230
		PO	*237.321*				Ti	237.196	80
		PO	*237.331*				*PO*	*237.199*	
		Cr	237.336	21 k			*PO*	*237.212*	
		PO	*237.350*				Mn	237.212	9600
		Fe	237.362	120			Ti	237.225	670
		W	237.447	9.3			Co	237.283	510
		Ti	237.459	180			Al	237.312	2.3
		W	237.476	34			Al	237.335	21
265.248	31	Ge	265.117	1.0			Fe	237.362	120
		Ge	265.157	3.1	226.374	140	Ti	226.178	610
		V	265.190	44			Co	226.260	680
		Ti	265.298	730			Al	226.346	5.7
		Ti	265.493	160			*NO*	*226.376*	
		Mo	265.503	660			W	**226.388**	12
257.540	53	V	257.402	32			Ti	226.402	17
		Co	257.435	64			V	226.440	380
		Al	257.509	5.7			Fe	226.505	980
		Mn	**257.551**	**75 k**					

6.2 Antimony (Sb)

The most sensitive analytical line for antimony is at 217.581 nm with a characteristic concentration of $c_0 = 0.20$ mg/L in an air/acetylene flame. The characteristic mass at this line, using a transversely heated graphite tube atomizer, is $m_0 = 21$ pg. No further information is available about this element at this point in time.

6.3 Arsenic (As)

The most sensitive analytical line for arsenic is at 193.696 nm at the beginning of vacuum-UV with a characteristic concentration of $c_0 = 0.6$ mg/L in an air/acetylene flame. The characteristic mass at this line, using a transversely heated graphite tube atomizer, is $m_0 = 50$ pg.

The air/acetylene flame already absorbs very strongly at the wavelength of arsenic. This kind of absorption due to flame gases is obviously absent in GF AAS. However, the arsenic line is also in the range of strong molecular absorption bands due to PO with pronounced fine structure, which can cause significant problems in GF AAS, and which may have to be corrected for in order to avoid interferences.

6.4 Barium (Ba)

The most sensitive analytical line for barium is at 553.548 nm with a characteristic concentration of $c_0 = 0.14$ mg/L in a nitrous oxide/acetylene flame. The characteristic mass at this line, using a transversely heated graphite tube atomizer, is $m_0 = 7$ pg.

A number of less sensitive analytical lines that might be used for the determination of higher barium concentrations are compiled in Table 6.3, together with information about their spectral environment.

Table 6.3: Selected absorption lines of barium (sorted according to their relative sensitivity) and corresponding spectral environment in order to identify species with potential risk for spectral interference, or additional elements that might be determined simultaneously

$\lambda_{element}$ / nm	Rel. sens.	Matrix name	λ / nm	Rel. sens.	$\lambda_{element}$ / nm	Rel. sens.	Matrix name	λ / nm	Rel. sens.
553.548	1.0				259.664	77	Ti	**259.659**	**26**
307.158	12	V	306.965	67	254.299	228	Fe	254.097	21
		Ti II	307.212				W	254.170	220
		Co	307.234	88			Ti	254.192	6.0
		Ti II	307.298				PO	**254.291**	
		V	307.382	26			PO	**254.298**	
350.111	13	Fe	349.784	270			V	254.373	420
		Ni	350.085	180			Co	254.425	31
		Co	350.228	30			W	254.534	11
		Co	350.262	170					

6.5 Beryllium (Be)

The most sensitive analytical line for beryllium is at 234.861 nm with a characteristic concentration of $c_0 = 0.006$ mg/L in a nitrous oxide / acetylene flame, which is about a factor of three better than in LS AAS. The characteristic mass at this line, using a transversely heated graphite tube atomizer, is $m_0 = 1.3$ pg, which is almost a factor of four better than in LS AAS.

The most sensitive analytical line for beryllium is within the range of intense NO molecular absorption bands, which might cause some problems in F AAS, and also within the range of PO bands, which might have to be considered in GF AAS in the presence of matrices containing high phosphate concentrations.

6.6 Bismuth (Bi)

The most sensitive analytical line for bismuth is at 223.061 nm with a characteristic concentration of $c_0 = 0.11$ mg/L in an air / acetylene flame, which is about a factor of two better than in LS AAS, and a linear working range up to about $A_{max} = 7$ mg/L. The characteristic mass at this line, using a transversely heated graphite tube atomizer, is $m_0 = 16$ pg.

A number of less sensitive analytical lines that might be used for the determination of higher bismuth concentrations are compiled in Table 6.4, together with information about their spectral environment.

The secondary bismuth absorption line at 306.772 nm is a doublet and directly overlaps with a rotational band of the OH molecule, as shown in Figure 6.3. Here the application of least-squares BC in F AAS is required in order to improve the SNR and to avoid potential interferences (refer to Section 8.1.1).

Table 6.4: Selected absorption lines of bismuth (sorted according to their relative sensitivity) and corresponding spectral environment in order to identify species with potential risk for spectral interference, or additional elements that might be determined simultaneously

$\lambda_{element}$ / nm	Rel. sens.	Matrix name	λ / nm	Rel. sens.	$\lambda_{element}$ / nm	Rel. sens.	Matrix name	λ / nm	Rel. sens.
223.061	1.0	V	222.884	620			Cu	223.008	300
		W	222.919	87			Ti	223.022	17
		Ti	222.973	39			V	223.037	330
		V	222.974	150			**Ti**	**223.048**	**23**

Table 6.4 (continued)

$\lambda_{element}$ / nm	Rel. sens.	Matrix name	λ / nm	Rel. sens.	$\lambda_{element}$ / nm	Rel. sens.	Matrix name	λ / nm	Rel. sens.
		In	**223.070**	**790**			**Cr II**	**206.158**	
		NO	*223.074*				**Tl**	**206.168**	**1200**
		V	223.142	320			**Ti**	**206.171**	**29**
		Sn	223.172	130			Ti	206.287	33
		Co	223.247	590			W	206.311	64
222.820	2.4	Ti	222.679	34			Ni	206.340	750
		Mo	222.739	78	227.658	14	Co	227.450	190
		Co	222.766	980			Ni	227.467	620
		Co	222.786	330			Sr	227.530	280
		W	222.798	20			Ca	227.547	60
		Fe	**222.817**	**1800**			Fe	227.603	55
		Co	**222.834**	**20 k**			**Co**	**227.653**	**59**
		V	222.884	620			**Ti**	**227.670**	**5.0**
		W	222.919	87			W	227.758	2.8
		Ti	222.973	39			In	227.820	420
		V	222.974	150			Ti	227.875	190
		Cu	223.008	300	211.022	18	Al	210.863	42
		Ti	223.022	17			Fe	210.896	170
		V	223.037	330			W	210.932	18
306.772	4.2	Pd	306.531	210			Mn	210.958	65
		Ti II	306.623				Al	210.998	65
		Ti II	306.635				**Fe**	**211.024**	**3800**
		V	306.637	2.6			W	211.114	220
		V	306.652	29			Ni	211.176	530
		OH	*306.768*		202.116	90	Pd	201.955	200
		OH	*306.776*				Pd	201.979	730
		OH	*306.792*				W	202.013	14
		V	306.965	67			Mo II	202.030	
206.163	7.6	Ni	205.990	96			Ti	202.094	110
		Ni	206.020	410			*NO*	*202.120*	
		Ti	206.059	33			**Ni**	**202.129**	**7300**
		Ag	206.117	150			W	202.184	220
		W	**206.152**	**160**			Pb	202.202	54
							Ti	202.218	310

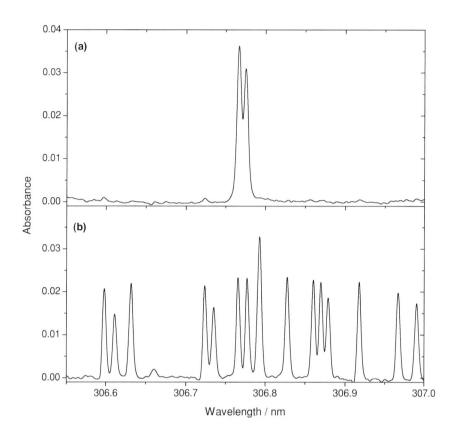

Figure 6.3: Secondary bismuth absorption line at 306.772 nm (a), and OH bands of the air / acetylene flame within the same spectral interval (b)

6.7 Boron (B)

The most sensitive analytical line for boron is at 249.772 nm with a characteristic concentration of about $c_0 = 4.4$ mg/L in a nitrous oxide / acetylene flame, which is almost a factor of three better than in LS AAS. The characteristic mass at this line, using a transversely heated graphite tube atomizer, is $m_0 = 76$ pg.

Boron actually has a second absorption line at 249.677 nm, i.e. within the spectral range of the detector, as shown in Figure 6.4, which might be used for the determination

6. The Individual Elements

of higher boron concentrations. However, as the sensitivity for boron is very low anyway in AAS, there is usually no need for a less sensitive analytical line.

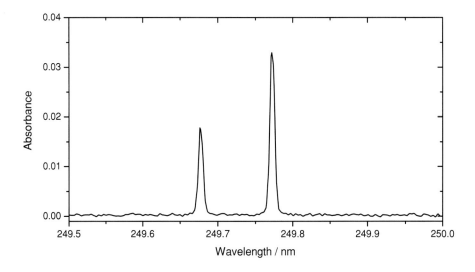

Figure 6.4: Two nearby absorption lines of boron at 249.677 nm and 249.772 nm, respectively

6.8 Cadmium (Cd)

Cadmium is an element of major environmental concern because of its high toxicity, and it has therefore to be determined in a wide variety of matrices at very low concentration levels.

The most sensitive analytical line for cadmium is at 228.802 nm with a characteristic concentration of $c_0 = 0.01$ mg/L in an air/acetylene flame, and a linear working range up to about $A_{max} = 0.4$ mg/L. The characteristic mass at this line, using a transversely heated graphite tube atomizer, is $m_0 = 0.8$ pg. The cadmium line is in the range of strong PO absorption bands, which have to be considered in GF AAS in the presence of high phosphate matrices. There is only one alternate line available for cadmium at 326.106 nm, which is about a factor of 150 less sensitive, and which might be used for the determination of very high cadmium concentrations.

Information about other atomic and molecular absorption lines that might be observed in the vicinity of the two cadmium lines is compiled in Table 6.4.

Table 6.5: Selected absorption lines of cadmium (sorted according to their relative sensitivity) and corresponding spectral environment in order to identify species with potential risk for spectral interference, or additional elements that might be determined simultaneously

$\lambda_{element}$ / nm	Rel. sens.	Matrix name	λ / nm	Rel. sens.	$\lambda_{element}$ / nm	Rel. sens.	Matrix name	λ / nm	Rel. sens.
228.802	1.0	Sn	228.668	35			Ni	228.840	240
		Fe	228.725	110			Ni	228.998	4.5
		Ni	228.733	370	326.106	150	In	325.856	13
		Co	228.781	110			Pd	325.878	250
		In	**228.802**	**13 k**			Sn	326.234	110
		PO	*228.806*				V	326.324	76

6.9 Calcium (Ca)

The most sensitive analytical line for calcium is at 422.673 nm with a characteristic concentration of $c_0 = 0.03$ mg/L in a nitrous oxide / acetylene flame. The characteristic mass at this line, using a transversely heated graphite tube atomizer, is $m_0 = 1.2$ pg. A few less sensitive analytical lines that might be used for the determination of higher calcium concentrations are listed in Table 6.6 together with information about their spectral environment.

There is an intense CN emission band head at 421.60 nm (refer to Section 7.2.6), which significantly deteriorates the SNR in LS AAS, unless a narrow slit width is used. In HR-CS AAS this interference obviously does not exist as this band head is more than 1 nm away from the calcium line.

6.10 Cesium (Cs)

The most sensitive analytical line for cesium is at 852.118 nm with a characteristic concentration of $c_0 = 0.065$ mg/L in an air / acetylene flame, and a linear working range up to about $A_{max} = 18$ mg/L. The characteristic mass at this line, using a transversely heated graphite tube atomizer, is $m_0 = 10$ pg. Cesium is significantly ionized in an air / acetylene flame, and requires the addition of another easily ionized element, such as potassium, in excess, in order to control this effect.

Table 6.6: Selected absorption lines of calcium (sorted according to their relative sensitivity) and corresponding spectral environment in order to identify species with potential risk for spectral interference, or additional elements that might be determined simultaneously

$\lambda_{element}$ / nm	Rel. sens.	Matrix name	λ / nm	Rel. sens.	$\lambda_{element}$ / nm	Rel. sens.	Matrix name	λ / nm	Rel. sens.
422.673	1.0						In	239.918	770
227.547	100	Ti	227.328	6.7			V	239.995	110
		Co	227.450	190	220.073	220	Ge	219.871	13
		Ni	227.467	620			W	219.892	11
		Sr	227.530	280			Al	219.918	570
		Cr	**227.531**	**21 k**			W	219.930	110
		Fe	**227.560**	**130 k**			Sn	219.934	5.2
		Fe	227.603	55			Cu	219.958	390
		Co	227.653	59			Ga	219.963	830
		Bi	227.658	14			V	220.018	460
		Ti	227.670	5.0			Fe	220.039	61
		W	227.758	2.8			**Ni**	**220.068**	**580**
239.856	140	W	239.773	8.0			**Fe**	**220.072**	**83**
		V	239.778	470			W	220.151	44
		W	239.798	5.9			Ni	220.156	300
		V	239.813	590			V	220.273	150
		V	239.827	890			W	220.286	60

6.11 Chromium (Cr)

Chromium can be determined in both a fuel-rich air/acetylene flame and a nitrous oxide/acetylene flame. The better sensitivity is attained in the former, but the determination is much more subject to interferences, so that the latter flame is preferred in the presence of complex matrices. The most sensitive analytical line for chromium is at 357.868 nm with a characteristic concentration of $c_0 = 0.040$ mg/L in an air/acetylene flame, and a linear working range up to about $A_{max} = 1$ mg/L. The characteristic mass at this line, using a transversely heated graphite tube atomizer, is $m_0 = 4$ pg. A number of less sensitive analytical lines that might be used for the determination of higher chromium concentrations are compiled in Table 6.7. The main chromium line at 357.868 nm is within the range of strong CN absorption bands (refer to Section 7.2.6), which might have to be considered when a nitrous oxide/acetylene flame is used for its determination.

Table 6.7: Selected absorption lines of chromium (sorted according to their relative sensitivity) and corresponding spectral environment in order to identify species with potential risk for spectral interference, or additional elements that might be determined simultaneously

$\lambda_{element}$ / nm	Rel. sens.	Matrix name	λ / nm	Rel. sens.	$\lambda_{element}$ / nm	Rel. sens.	Matrix name	λ / nm	Rel. sens.
357.868	1.0	Co	357.536	61			Cr	236.681	150
		PO	*357.876*				Al	236.705	4.1
		Fe	358.120	98			W	236.769	18
359.348	1.4	V	359.026	720	236.681	150	Cr	236.473	82
		Co	359.486	150			Co	236.506	51
							W	236.545	15
360.532	1.6	Co	360.208	200			Cr	236.591	120
		Ti	360.428	360			W	236.618	66
		Co	**360.537**	**1400**			*NO*	*236.675*	
		W	360.606	70			*PO*	*236.690*	
		Ti	360.679	500			Al	236.705	4.1
		Fe	360.886	920			W	236.769	18
425.433	3.0						Ti	236.856	400
428.972	4.6	Ti	428.601	69	301.492	160	Ni	301.200	35
		Ti	428.741	72			Co	301.359	340
		Ti	428.907	59			Cr	301.370	640
		Ti	429.093	74			W	301.379	43
302.156	73	Ni	301.914	100			**Cr**	**301.476**	**350**
		Fe	302.049	42			Cr	301.520	380
		Fe	302.064	3.9			W	301.647	41
		Cr	302.067	610			W	301.744	8.1
		Fe	302.107	11			Co	301.754	130
		OH	*302.176*				Cr	301.757	130
		Fe	302.403	180			Fe	301.763	200
		Cr	302.433	320	298.647	190	Fe	298.357	22
236.591	120	Cr	236.473	82			Ni	298.413	160
		Co	236.506	51			Ti	298.548	380
		W	236.545	15			Cr	298.600	380
		PO	*236.577*				*OH*	*298.626*	
		NO	*236.596*				**Fe**	**298.646**	**11 k**
		PO	*236.605*				Co	298.716	96
		W	236.618	66			Cr	298.865	900

6.12 Cobalt (Co)

The most sensitive analytical line for cobalt is at 240.725 nm with a characteristic concentration of $c_0 = 0.034$ mg/L in an air/acetylene flame, which is about a factor of two higher sensitivity than with LS AAS, and a linear working range up to about $A_{max} = 1.3$ mg/L. The characteristic mass at this line, using a transversely heated graphite tube atomizer, is $m_0 = 7$ pg. Cobalt has more than 500 absorption lines of significantly different sensitivity. A selection of less sensitive analytical lines that might be used for the determination of higher cobalt concentrations is given in Table 6.8, together with information about their spectral environment.

Table 6.8: Selected absorption lines of cobalt (sorted according to their relative sensitivity) and corresponding spectral environment in order to identify species with potential risk for spectral interference, or additional elements that might be determined simultaneously

$\lambda_{element}$ / nm	Rel. sens.	Matrix name	λ / nm	Rel. sens.	$\lambda_{element}$ / nm	Rel. sens.	Matrix name	λ / nm	Rel. sens.
240.725	1.0	W	240.558	2.3	252.136	2.7	Ti	251.902	63
		W	240.569	9.9			Si	251.920	4.3
		W	240.618	23			V	251.963	31
		V	240.675	85			Ti	252.054	13
		V	240.790	140			PO	*252.121*	
		W	240.904	100			W	**252.132**	**14**
		W	240.914	170			In	**252.137**	**110**
241.162	2.1	Ti	241.137	180			PO	*252.153*	
		Ti	**241.158**	**180**			W	252.216	130
		Pb	**241.173**	**9500**			Fe	252.285	1.7
		Ni	241.264	400			In	252.298	710
		V	241.269	66			W	252.341	9.7
		Co	241.276	49	243.221	4.2	W	242.985	70
		V	241.304	91			In	242.986	870
		W	241.378	180			W	243.108	8.5
242.493	2.2	V	242.337	280			V	243.157	750
		Ni	242.403	750			V	243.195	210
		W	242.421	8.3			V	243.202	190
		Ti	242.425	15			PO	*243.236*	
		PO	*242.492*				Ti	243.322	59
		V	242.613	940			W	243.398	10
							Ti	243.410	140

Table 6.8 (continued)

$\lambda_{element}$ / nm	Rel. sens.	Matrix name	λ / nm	Rel. sens.	$\lambda_{element}$ / nm	Rel. sens.	Matrix name	λ / nm	Rel. sens.
241.446	4.7	Ni	241.264	400	243.666	8.9	V	243.552	57
		V	241.269	66			W	243.579	68
		Co	241.276	49			Co	243.582	480
		V	241.304	91			W	243.596	2.3
		W	241.378	180			W	243.626	80
		W	241.404	9.7			*PO*	*243.659*	
		Co	241.529	7.3			**Na**	**243.660**	**1100 k**
		V	241.533	68			**W**	**243.662**	**17**
		W	241.568	3.0			Ti	243.829	330
		W	241.624	200			Co	243.904	20
		V	241.675	32	230.901	14	Sr	230.730	130
252.897	5.6	Fe	252.744	4.8			Ni	230.736	320
		Ti	252.798	63			*PO*	*230.889*	
		W	252.848	42			**Ti**	**230.890**	**130**
		Si	252.851	3.0			**Fe**	**230.900**	**180**
		PO	*252.902*				**W**	**230.902**	**22**
		Fe	**252.914**	**12**			*PO*	*230.903*	
		Fe	252.984	74			V	231.017	940
		Ti	252.986	7.8			Ni	231.096	1.7
		Co	253.013	290	350.228	30	Ni	350.085	180
		V	253.018	40			Ba	350.111	36
		Ti II	253.126				Co	350.262	170
241.529	7.3	V	241.304	91	340.917	100	Co	341.233	47
		W	241.378	180					
		W	241.404	9.7	344.917	100	Ni	344.626	22
		Co	241.446	4.7			Mo	344.712	360
		V	**241.533**	**68**			Co	344.944	130
		W	241.568	3.0	341.716	240	Ni	341.394	380
		W	241.624	200			Ni	341.476	4.8
		V	241.675	32					
		V	241.735	55					
		Ge	241.737	83					

6.13 Copper (Cu)

The most sensitive analytical line for copper is at 324.754 nm with a characteristic concentration of $c_0 = 0.018$ mg/L in an air/acetylene flame, which is about a factor of two higher sensitivity than with LS AAS, and a linear working range up to about $A_{max} = 2.4$ mg/L. The characteristic mass at this line, using a transversely heated graphite tube atomizer, is about $m_0 = 6$ pg. The main resonance line of copper is within an absorption band of the PO molecule with a band head at 324.62 nm (refer to Section 7.2.14), which has to be considered in the presence of high phosphate matrices, mainly in GF AAS.

A number of less sensitive analytical lines that might be used for the determination of higher copper concentrations are compiled in Table 6.9, together with information about their spectral environment.

Table 6.9: Selected absorption lines of copper (sorted according to their relative sensitivity) and corresponding spectral environment in order to identify species with potential risk for spectral interference, or additional elements that might be determined simultaneously

$\lambda_{element}$ / nm	Rel. sens.	Matrix name	λ / nm	Rel. sens.	$\lambda_{element}$ / nm	Rel. sens.	Matrix name	λ / nm	Rel. sens.
324.754	1.0	La II	324.513		216.509	8.0	W	216.301	62
		La	324.702	11			Co	216.357	260
		Ni	324.846	790			Fe	216.386	870
		V	324.956	570			W	216.420	62
327.396	2.2	V	327.163	85			W	216.434	13
							Fe	216.455	270
217.894	5.6	Fe	217.684	720			Al	216.458	980
		V	217.701	25			W	216.493	43
		Bi	217.721	620			Ni	216.615	570
		V	217.724	370			Fe	216.677	3.5
		Fe	217.812	8.0	218.172	9.4	W	217.977	11
		Pd	217.828	940			In	217.990	990
		W	217.849	14			Co	218.006	180
		W	217.977	11			W II	218.079	
		In	217.990	990			V	218.198	330
		Co	218.006	180			V	218.222	24
		W II	218.079				Ni	218.238	290
							W	218.290	20

Table 6.9 (continued)

$\lambda_{element}$ / nm	Rel. sens.	Matrix name	λ / nm	Rel. sens.	$\lambda_{element}$ / nm	Rel. sens.	Matrix name	λ / nm	Rel. sens.
222.570	20	W	222.358	16			NO	202.422	
		Mo	222.392	110			Ni	202.539	230
		Ti	222.511	18			Mg	202.582	23
		Co	222.535	320	249.215	92	W	248.972	7.8
		V	222.543	240			Fe	248.975	12
		Mo	222.543	81			Fe	249.064	2.7
		W	222.554	33			Fe	249.116	5.9
		V	**222.580**	**990**			**OH**	**249.228**	
		Co	**222.585**	**46 k**			W	249.237	83
		W II	222.589						
		Ti	222.679	34	224.426	260	W	224.206	42
		Mo	222.739	78			**NO**	**224.416**	
		Co	222.766	980			Ni	224.448	110
		Co	222.786	330			Ni	224.453	320
202.434	38	W	202.390	130			Ti	224.469	98
		Ti	202.403	71			V	224.576	490
							Sn	224.605	1.0

6.14 Europium (Eu)

The most sensitive analytical line for europium is at 459.403 nm with a characteristic concentration of $c_0 = 0.15$ mg/L in a nitrous oxide / acetylene flame. The characteristic mass at this line, using a transversely heated graphite tube atomizer, is $m_0 = 9$ pg. No further information is available about this element.

6.15 Gallium (Ga)

The most sensitive analytical line for gallium is at 287.424 nm with a characteristic concentration of $c_0 = 0.94$ mg/L in an air / acetylene flame, and a linear working range up to about $A_{max} = 80$ mg/L. The characteristic mass at this line, using a transversely heated graphite tube atomizer, is $m_0 = 20$ pg. The main line for gallium is within the range of an OH absorption band, which should be considered in F AAS. A number of less sensitive analytical lines that might be used for the determination of higher gallium concentrations are given in Table 6.10, together with information about their spectral environment.

Table 6.10: Selected absorption lines of gallium (sorted according to their relative sensitivity) and corresponding spectral environment in order to identify species with potential risk for spectral interference, or additional elements that might be determined simultaneously

$\lambda_{element}$ / nm	Rel. sens.	Matrix name	λ / nm	Rel. sens.	$\lambda_{element}$ / nm	Rel. sens.	Matrix name	λ / nm	Rel. sens.
287.424	1.0	**Fe**	**287.417**	**630**			Fe	250.113	10
294.364	1.0	Fe	294.134	280			V	250.161	76
		Ti	294.199	2.4			W	250.178	72
		V	294.233	75	245.008	8	Fe	244.771	200
		V	294.320	43			Pd	244.791	1.0
		Ni	294.391	70			W	244.839	170
		Ga	294.417	8.9			*PO*	*245.007*	
		Mo	294.421	350			*OH*	*245.009*	
		W	294.440	2.3			*OH*	*245.016*	
		V	294.653	110			W	245.135	83
403.299	1.2	Mn	403.075	15			W II	245.148	
		Mn	**403.306**	**120**			W	245.200	1.9
		Mn	403.448	150	294.417	9	Fe	294.134	280
417.204	1.5						Ti	294.199	2.4
							V	294.233	75
250.019	6.7	B	249.772	1.0			V	294.320	43
		Ge	249.796	23			Ga	294.364	1.0
		V	249.804	940			Ni	294.391	70
		V	249.824	990			**Mo**	**294.421**	**350**
		V	249.911	820			W	294.440	2.3
		Ba	**250.036**	**2300**			V	294.653	110
		Ga	250.071	66			W	294.699	2.5

6.16 Germanium (Ge)

The most sensitive analytical line for germanium is at 265.157 nm with a characteristic concentration of $c_0 = 0.85$ mg/L in a nitrous oxide/acetylene flame. The characteristic mass at this line, using a transversely heated graphite tube atomizer, is about $m_0 = 55$ pg. Germanium has almost 30 analytical lines with varying sensitivity; however, no further information is given here, as the sensitivity at the main analytical line is not very high anyway, and there is usually no need to reduce the sensitivity of the determination further.

6.17 Gold (Au)

The most sensitive analytical line for gold is at 242.795 nm with a characteristic concentration of $c_0 = 0.06$ mg/L in an oxidizing air/acetylene flame, and a linear working range up to about $A_{max} = 3.3$ mg/L. The characteristic mass at this line, using a transversely heated graphite tube atomizer, is $m_0 = 11$ pg. As a less sensitive analytical line the one at 267.595 nm may be used for the determination of higher gold concentrations.

6.18 Indium (In)

The most sensitive analytical line for indium is the doublet line at 303.936 nm with a separation of 4.2 pm between its components which becomes visible only by using an extremely high-resolution spectrometer like the SuperDEMON (see Figure 6.5). The line shows a characteristic concentration of $c_0 = 0.15$ mg/L in an oxidizing air/acetylene flame, and a linear working range up to about $A_{max} = 33$ mg/L. The characteristic mass at this line, using a transversely heated graphite tube atomizer, is $m_0 = 12$ pg. Indium has a relatively large number of less sensitive analytical lines that might be used for the determination of higher analyte concentrations. However, as there is only rarely a need to reduce the sensitivity for this element significantly, only the most sensitive absorption lines are listed in Table 6.11, together with information about their spectral environment.

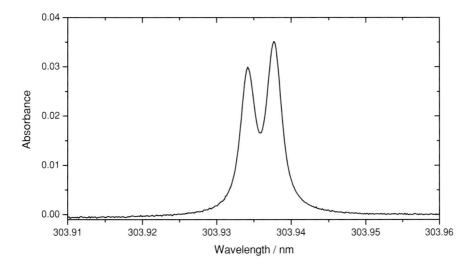

Figure 6.5: Indium doublet line at 303.936 nm, measured with the SuperDEMON spectrometer

6. The Individual Elements

Table 6.11: Selected absorption lines of indium (sorted according to their relative sensitivity) and corresponding spectral environment in order to identify species with potential risk for spectral interference, or additional elements that might be determined simultaneously

$\lambda_{element}$ / nm	Rel. sens.	Matrix name	λ / nm	Rel. sens.	$\lambda_{element}$ / nm	Rel. sens.	Matrix name	λ / nm	Rel. sens.
303.936	1.0	Cr	303.704	910	451.131	4.0	Ti	451.273	81
		Fe	303.739	28	256.015	8	V	255.890	570
		Ni	303.793	20			W	256.197	12
		Ge	303.907	17			Co	256.212	91
		Cr	304.084	970			V	256.213	46
		W	304.186	33			Li	256.231	600
325.609	1.8	Mn	325.613	290 k	325.856	13	In	325.609	1.8
		In	325.856	13			Mn	325.841	270 k
		Pd	325.878	250			Pd	325.878	250
410.176	2.9	V	409.978	28			Cd	326.106	150
		W	410.270	41					
		V	410.516	30					

6.19 Iridium (Ir)

The most sensitive analytical line for iridium is at 208.882 nm with a characteristic concentration of about $c_0 = 3.2$ mg/L in a reducing air / acetylene flame, and a linear working range up to about 55 mg/L. Iridium may also be determined in a nitrous oxide / acetylene flame with about a factor of two lower sensitivity, but significantly reduced interferences. The characteristic mass at this line, using a transversely heated graphite tube atomizer, is $m_0 = 120$ pg. No further information is available about the determination of this element.

6.20 Iron (Fe)

Iron is the most abundant element on our planet, and the Earth's crust contains about 5 % of this metal, so that contamination is often a major issue in its determination. The most sensitive analytical line for iron is at 248.327 nm with a characteristic concentration of $c_0 = 0.044$ mg/L in an oxidizing air / acetylene flame, and a linear working range up to about $A_{max} = 1.8$ mg/L. The characteristic mass at this line, using a transversely heated graphite tube atomizer, is $m_0 = 8$ pg. Iron has more than 500 absorption lines with greatly

6.20 Iron (Fe)

differing sensitivity, so that only a small fraction of these lines, covering a sensitivity range of more than two orders of magnitude, had to be selected for presentation in Table 6.12, together with information about their spectral environment.

Table 6.12: Selected absorption lines of iron (sorted according to their relative sensitivity) and corresponding spectral environment in order to identify species with potential risk for spectral interference, or additional elements that might be determined simultaneously

$\lambda_{element}$ / nm	Rel. sens.	Matrix name	λ / nm	Rel. sens.	$\lambda_{element}$ / nm	Rel. sens.	Matrix name	λ / nm	Rel. sens.
248.327	1.0	W	248.096	8.2	249.064	2.7	W	248.891	160
		W	248.144	4.8			Ni	248.951	920
		V	248.212	480			W	248.972	7.8
		W	248.221	160			Fe	248.975	12
		V	248.274	690			**Na**	**249.073**	**100 k**
		V	248.309	940			Fe	249.116	5.9
		V	**248.337**	**470**			Cu	249.215	92
		Sn	**248.339**	**39**			W	249.237	83
		Co	248.361	490	271.903	3.0	**W**	**271.891**	**3.5**
		Fe	248.419	13			**Mn**	**271.902**	**240 k**
		W	248.474	72			Ga	271.966	20
248.814	1.5	Fe	248.637	210			W	271.986	53
		W	248.750	20			Fe	272.090	8.6
		V	**248.820**	**1300**	216.677	3.5	W	216.493	43
		W	248.891	160			Cu	216.509	8.0
		Ni	248.951	920			Ni	216.615	570
		W	248.972	7.8			W	216.769	140
		Fe	248.975	12			W	216.830	46
252.285	1.7	Ti	252.054	13			Tl	216.855	44
		W	252.132	14			Co	216.871	520
		Co	252.136	2.7			Al	216.883	18
		In	252.137	110	302.064	3.9	Fe	301.763	200
		W	252.216	130			Cr	301.849	890
		PO	*252.268*				Ni	301.914	100
		PO	*252.293*				**Fe**	**302.049**	**42**
		In	252.298	710			**Cr**	**302.067**	**610**
		W	252.341	9.7			Fe	302.107	11
		Si	252.411	2.9			Cr	302.156	73
		Fe	252.429	26					

Table 6.12 (continued)

$\lambda_{element}$ / nm	Rel. sens.	Matrix name	λ / nm	Rel. sens.	$\lambda_{element}$ / nm	Rel. sens.	Matrix name	λ / nm	Rel. sens.
252.744	4.8	Ti II	252.561				*PO*	*251.083*	
		W	252.568	60			Co	251.102	47
		V	252.622	16			V	251.165	89
		W	252.642	170			V	251.195	52
		PO	*252.746*				Fe	251.236	350
		Ti	252.798	63	344.061	26	Ni	343.728	66
		W	252.848	42			Fe	344.099	52
		Si	252.851	3.0			Pd	344.140	190
		Co	252.897	5.6			Co	344.292	110
		Fe	252.914	12			Co	344.364	61
		Fe	252.984	74			Fe	344.388	170
		Ti	252.986	7.8	344.099	52	Fe	344.061	26
249.116	5.9	W	248.891	160			Pd	344.140	190
		Ni	248.951	920			Co	344.292	110
		W	248.972	7.8			Co	344.364	61
		Fe	248.975	12			Fe	344.388	170
		Fe	249.064	2.7					
		OH	*249.114*		388.628	100	Ba	388.933	790
		Cu	249.215	92			Ti	388.996	330
		W	249.237	83	373.332	210	Ti	372.981	3.4
251.084	9.4	W	250.874	64			Fe	373.486	130

6.21 Lanthanum (La)

The elements of Group III of the periodic system with atomic numbers from 57 to 71, i.e. La – Lu, are collectively referred to as the lanthanoids. Together with scandium and yttrium, the lanthanoids comprise the rare-earth elements.

The most sensitive analytical line for lanthanum is at 550.134 nm with a characteristic concentration of $c_0 = 45$ mg/L in a nitrous oxide/acetylene flame. The characteristic mass at this line, using a transversely heated graphite tube atomizer, is $m_0 = 7000$ pg. Lanthanum has more than 30 analytical lines with varying sensitivity; however, no further information is given here, as the sensitivity at the main analytical line is not very high anyway, and there is usually no need to reduce the sensitivity of the determination further.

6.22 Lead (Pb)

Lead is among the elements most frequently determined by AAS, mostly because of its toxicity, and its abundant use as an antiknock agent in gasoline, as paint and for many other industrial purposes. Although many of these applications have been abandoned in the meantime, there is still a significant environmental burden left from previous decades that calls for regular monitoring.

The most sensitive analytical line for lead is at 217.001 nm with a characteristic concentration of $c_0 = 0.073$ mg/L in an air / acetylene flame. In contrast to LS AAS, where this line exhibits a poor SNR and a pronounced non-linearity, it can be used without problems in HR-CS AAS because of the much higher radiation intensity available from the xenon short-arc lamp. The characteristic mass at this line, using a transversely heated graphite tube atomizer, is $m_0 = 6$ pg. The absorption at 217.001 nm is in the vicinity of PO absorption bands, which might have to be considered in GF AAS in the presence of phosphate matrix. A number of less sensitive analytical lines that might be used for the determination of higher lead concentrations are listed in Table 6.13 together with information about their spectral environment.

Table 6.13: Selected absorption lines of lead (sorted according to their relative sensitivity) and corresponding spectral environment in order to identify species with potential risk for spectral interference, or additional elements that might be determined simultaneously

$\lambda_{element}$ / nm	Rel. sens.	Matrix name	λ / nm	Rel. sens.	$\lambda_{element}$ / nm	Rel. sens.	Matrix name	λ / nm	Rel. sens.
217.001	1.0	W	216.830	46	205.328	29	Ni	205.207	140
		Tl	216.855	44			W II	205.313	
		Co	216.871	520			**PO**	**205.342**	
		Al	216.883	18			Ge	205.446	32
		W	216.948	6.2			W II	205.460	
		Al	216.984	530	261.418	47	W	261.308	6.0
		Fe	**216.994**	**61 k**			W	261.382	7.8
		W	217.020	87			**Co**	**261.412**	**2200**
		V	217.074	36			W	261.512	57
		Fe	217.130	300			Mo	261.678	910
283.306	2.4	W	283.138	3.6	202.202	54	W	202.013	14
		W	283.363	18			Mo II	202.030	
		W	283.564	42			Ti	202.094	110

Table 6.13 (continued)

$\lambda_{element}$ / nm	Rel. sens.	Matrix name	λ / nm	Rel. sens.	$\lambda_{element}$ / nm	Rel. sens.	Matrix name	λ / nm	Rel. sens.
		Bi	202.115	90	280.200	200	W	279.993	19
		W	202.184	220			Mn	280.108	2.0
		NO	*202.198*				Ti	280.250	32
		Ti	202.218	310			Mg II	280.270	
		W	202.390	130	368.346	210	Fe	367.991	170
		Ti	202.403	71			W	368.208	140
							V	368.312	100

6.23 Lithium (Li)

The most sensitive analytical line for lithium is at 670.785 nm with a characteristic concentration of $c_0 = 0.006$ mg/L in an air/acetylene flame, and a linear working range up to about $A_{max} = 3$ mg/L. The characteristic mass at this line, using a transversely heated graphite tube atomizer, is $m_0 = 1$ pg. The lithium line at 670.785 nm is actually a doublet with a separation of about 15 pm, and it also exhibits an isotope shift of the same magnitude [150]. However line broadening in conventional atomizers is usually too pronounced, so that the individual fine structure components cannot be resolved.

Lithium has one less sensitive analytical line that may be used for the determination of higher analyte concentrations. Both absorption lines are compiled in Table 6.14, together with information about their spectral environment.

Table 6.14: Selected absorption lines of lithium (sorted according to their relative sensitivity) and corresponding spectral environment in order to identify species with potential risk for spectral interference, or additional elements that might be determined simultaneously

$\lambda_{element}$ / nm	Rel. sens.	Matrix name	λ / nm	Rel. sens.	$\lambda_{element}$ / nm	Rel. sens.	Matrix name	λ / nm	Rel. sens.
670.785	1.0						*PO*	*323.283*	
323.266	250	V	323.064	670			Ni	323.293	54
		PO	*323.242*				Ti II	323.452	
							Ni	323.465	370

6.24 Magnesium (Mg)

Magnesium is one of the elements most frequently determined by FAAS, which in part can be ascribed to the very high sensitivity that can be attained for this element. The most sensitive analytical line for magnesium is at 285.213 nm with a characteristic concentration of $c_0 = 0.003$ mg/L in an air/acetylene flame, and a linear working range up to about $A_{max} = 0.25$ mg/L. The characteristic mass at this line, using a transversely heated graphite tube atomizer, is $m_0 = 0.5$ pg. The main resonance line at 285.213 nm is in the range of strong OH absorption bands, which might have to be considered when this element is determined by FAAS. Magnesium has one less sensitive analytical line that may be used for the determination of higher analyte concentrations. The two lines are shown in Table 6.15, together with information about their spectral environment.

Table 6.15: Selected absorption lines of magnesium (sorted according to their relative sensitivity) and corresponding spectral environment in order to identify species with potential risk for spectral interference, or additional elements that might be determined simultaneously

$\lambda_{element}$ / nm	Rel. sens.	Matrix name	λ / nm	Rel. sens.	$\lambda_{element}$ / nm	Rel. sens.	Matrix name	λ / nm	Rel. sens.
285.213	1.0	V	285.175	200			Ni	202.539	230
202.582	23	W	202.390	130			Ni	202.659	73
		Ti	202.403	71			Sn	202.698	170
		Cu	202.434	38			V	202.764	200

6.25 Manganese (Mn)

Manganese has a line triplet, as shown in Figure 6.6, and all three lines are within the spectral window of HR-CS AAS, and may be used simultaneously for analytical purposes. The most sensitive analytical line for manganese is at 279.482 nm with a characteristic concentration of $c_0 = 0.012$ mg/L in an air/acetylene flame. The characteristic mass at this line, using a transversely heated graphite tube atomizer, is $m_0 = 1.8$ pg. The main resonance line for manganese at 279.482 nm is in the range of strong OH absorption bands, which might have to be considered when this element is determined by FAAS.

Manganese has a lot of absorption lines; a selection of less sensitive analytical lines that might be used for the determination of higher manganese concentrations are compiled in Table 6.15, together with information about their spectral environment.

6. The Individual Elements

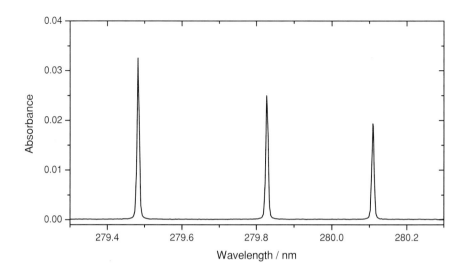

Figure 6.6: Manganese triplet with absorption lines at 279.482 nm, 279.827 nm and 280.108 nm, respectively

Table 6.16: Selected absorption lines of manganese (sorted according to their relative sensitivity) and corresponding spectral environment in order to identify species with potential risk for spectr. interference, or additional elements that might be determined simultaneously

$\lambda_{element}$ / nm	Rel. sens.	Matrix name	Matrix λ / nm	Rel. sens.	$\lambda_{element}$ / nm	Rel. sens.	Matrix name	Matrix λ / nm	Rel. sens.
279.482	1.0	W	279.270	10			Ti	280.250	32
		Fe	**279.470**	**83 k**			Mg II	280.270	
		Fe	**279.500**	**5600**	403.075	15	Ga	403.299	1.2
		Mg II	279.553				Mn	403.306	120
279.827	1.5	Mg II	279.553				Mn	403.448	150
		Ni	279.865	310	200.385	22	W II	200.247	
		W	279.993	19			Ti	200.357	800
280.108	2.0	Ni	279.865	310			**Fe**	**200.387**	**8400**
		W	279.993	19			Ti	200.414	670
		Pb	280.200	200					

6.25 Manganese (Mn)

Table 6.16 (continued)

$\lambda_{element}$ / nm	Rel. sens.	Matrix name	λ / nm	Rel. sens.	$\lambda_{element}$ / nm	Rel. sens.	Matrix name	λ / nm	Rel. sens.
222.184	28	Ti	221.975	200	209.216	69	Al	209.013	240
		Mo	222.025	36			Fe	209.038	600
		Ti	222.145	17			W	209.048	5.9
		W	**222.185**	**26**			V	209.054	360
		Ni	**222.195**	**540**			V	209.069	21
		V	222.284	350			V	209.090	120
		V	222.302	150			V	209.131	28
		Ti	222.319	32			Sn	209.158	11
		W	222.358	16			Al	209.169	190
		Mo	222.392	110			Al	209.191	170
221.385	42	Si	221.174	17			V	209.191	590
		Ni	221.216	920			V	209.234	23
		V	**221.370**	**990**			V	209.244	7.2
		Co	**221.390**	**150**			Mn	209.250	180
		Ga	221.440	700			W	209.254	48
		Cu	221.458	970			Pd	209.263	810
		W	221.480	39			Mn	209.340	90
210.958	55	Al	210.863	42			Fe	209.368	66
		Fe	210.896	170			W	209.372	80
		W	210.932	18			Al	209.407	130
		Al	210.998	65			W	209.408	120
		Bi	211.022	18	209.340	90	V	209.131	28
		W	211.114	220			Sn	209.158	11
220.881	67	Tl	220.695	71			Al	209.169	190
		Co	220.786	260			Al	209.191	170
		Si	220.798	14			V	209.191	590
		Ni	**220.867**	**52 k**			Mn	209.216	69
		Fe	**220.872**	**65 k**			V	209.234	23
		W	220.906	180			V	209.244	7.2
		Sn	220.965	7.6			Mn	209.250	180
		Al	221.006	5.8			W	209.254	48
		Tl	221.065	140			Pd	209.263	810
		Tl	221.079	300			**Co**	**209.340**	**3500**
		Si	221.089	5.9			Fe	209.368	66
		Ni	221.101	760			W	209.372	80
							Al	209.407	130

Table 6.16 (continued)

$\lambda_{element}$ / nm	Rel. sens.	Matrix name	λ / nm	Rel. sens.	$\lambda_{element}$ / nm	Rel. sens.	Matrix name	λ / nm	Rel. sens.
		W	209.408	120			La	403.721	6.7
		Ge	209.426	1.3	209.250	180	W	209.048	5.9
		Sn	209.435	94			V	209.054	360
		V	209.471	38			V	209.069	21
		W II	209.475				V	209.090	120
		Al	209.483	150			V	209.131	28
210.607	120	V	210.456	420			Sn	209.158	11
		Tl	210.460	170			Al	209.169	190
		Co	210.473	780			Al	209.191	170
		V	210.483	420			V	209.191	590
		W	210.540	10			Mn	209.216	69
		Ge	210.582	25			V	209.234	23
		PO	*210.598*				**V**	**209.244**	**7.2**
		V	210.631	600			**W**	**209.254**	**48**
		Fe	210.640	110			**Pd**	**209.263**	**810**
		Co	210.680	460			Mn	209.340	90
403.306	120	Mn	403.075	15			Fe	209.368	66
		Ga	**403.299**	**1.2**			W	209.372	80
		Mn	403.448	150			Al	209.407	130
403.448	150	Mn	403.075	15			W	209.408	120
		Ga	403.299	1.2			Ge	209.426	1.3
		Mn	403.306	120			Sn	209.435	94

6.26 Mercury (Hg)

Mercury is, without doubt, the most important element nowadays in analytical chemistry, as it has been declared a global pollutant because of the extreme toxicity of mercury vapor and many of the mercury compounds, particularly its methylated species.

The most sensitive analytical line for mercury at 184.889 nm is not accessible with normal AA spectrometers, including HR-CS AAS. The only absorption line that is available is that at 253.652 nm, which originates from a metastable state. As mercury usually has to be determined at extremely low trace levels, F AAS is not suitable for this purpose. GF AAS might be used successfully for the determination of mercury in environmental samples that are difficult to be brought into solution, using direct solid sample analysis. The characteristic mass, using a transversely heated graphite tube atomizer, is $m_0 = 160$ pg.

The most sensitive technique for solution analysis is cold vapor AAS or AFS using collection of the mercury vapor on a gold trap or directly in the graphite tube, treated with a permanent modifier.

6.27 Molybdenum (Mo)

The most sensitive analytical line for molybdenum is at 313.259 nm with a characteristic concentration of $c_0 = 0.39$ mg/L in a nitrous oxide/acetylene flame. The characteristic mass at this line, using a transversely heated graphite tube atomizer, is $m_0 = 13$ pg. Molybdenum has about 40 less sensitive analytical lines that might be used for the determination of higher analyte concentrations. However, as the sensitivity for molybdenum is not very high anyway, there is only very rarely a need to reduce the sensitivity further, so that only a few selected lines are given in Table 6.17, together with information about their spectral environment.

Table 6.17: Selected absorption lines of molybdenum (sorted according to their relative sensitivity) and corresponding spectral environment in order to identify species with potential risk for spectral interference, or additional elements that might be determined simultaneously

$\lambda_{element}$ / nm	Rel. sens.	Matrix name	λ / nm	Rel. sens.	$\lambda_{element}$ / nm	Rel. sens.	Matrix name	λ / nm	Rel. sens.
313.259	1.0	Ni	313.410	21			Ti	390.096	58
317.035	1.7	Ti II	316.853				V	390.225	19
							CN	*390.292*	
319.397	1.8	W	319.157	18			**Fe**	**390.295**	**28 k**
		Ti	319.200	1.0			Ti	390.478	31
		Ba	**319.391**	**23 k**	320.883	6.8	Ti	320.584	250
386.411	2.2	*CN*	*386.430*				W	320.725	35
		V	386.486	22			V	320.741	22
390.296	3.1	Fe	389.971	330			W	320.828	47

6.28 Nickel (Ni)

The most sensitive analytical line for nickel is at 232.003 nm with a characteristic concentration of $c_0 = 0.04$ mg/L in an air/acetylene flame. The characteristic mass at this line, using a transversely heated graphite tube atomizer, is $m_0 = 10$ pg. The main absorption line of nickel is in the range of strong PO absorption bands, which might have to be considered in GF AAS analysis in the presence of phosphate matrix.

There are two secondary nickel lines at 232.138 nm and 232.195 nm respectively with relative sensitivities of about 1:7 and 1:50 in the vicinity of the main analytical line, which is within the spectral range of HR-CS AAS, as shown in Figure 6.7, and which can be used simultaneously with the primary line. In this way three calibration curves can be established using a single measurement, thus extending the linear working range by at least two orders of magnitude without the need for additional measurements.

Nickel has about 350 absorption lines with greatly differing sensitivity, so that only a small fraction of these lines, covering a sensitivity range of more than two orders of magnitude, had to be selected for presentation in Table 6.18, together with information about their spectral environment.

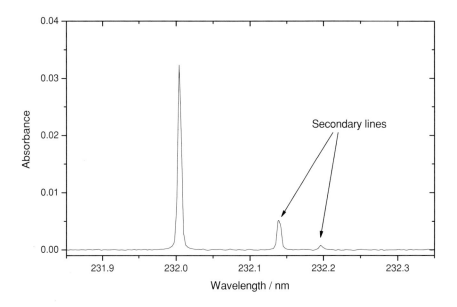

Figure 6.7: Most sensitive nickel absorption line at 232.003 nm and two secondary lines at 232.138 nm and 232.195 nm, respectively

6.28 Nickel (Ni)

Table 6.18: Selected absorption lines of nickel (sorted according to their relative sensitivity) and corresponding spectral environment in order to identify species with potential risk for spectral interference, or additional elements that might be determined simultaneously

$\lambda_{element}$ / nm	Rel. sens.	Matrix name	λ / nm	Rel. sens.	$\lambda_{element}$ / nm	Rel. sens.	Matrix name	λ / nm	Rel. sens.
232.003	1.0	W	231.858	140			*PO*	*231.251*	
		V	**232.014**	**300**			Fe	231.310	110
		Fe	232.036	74			W	231.317	3.1
		Ni	232.138	7.2			Ni	231.366	5.9
		W	232.163	6.2			Ni	231.398	10
		Ni	232.195	49			W	231.417	42
231.096	1.7	Ti	230.890	130			Ge	231.420	260
		Fe	230.900	180			Ti	231.430	150
		Co	230.901	14	228.998	4.5	Co	228.781	110
		W	230.902	22			Cd	228.802	1.0
		V	231.017	940			Ni	228.840	240
		NO	*231.084*				*PO*	*228.983*	
		PO	*231.102*				*PO*	*228.999*	
		Co	231.136	930			**Fe**	**229.007**	**36 k**
		V	231.146	650			*PO*	*229.013*	
		Ni	231.234	4.4			**In**	**229.013**	**5600**
		Fe	231.310	110			W	229.095	14
		W	231.317	3.1			Co	229.145	380
234.554	3.5	W	234.374	110	232.580	4.5	V	232.454	820
		NO	*234.548*				V	232.474	180
		W	234.600	110			Co	232.554	150
		Co	234.616	130			**V**	**232.586**	**340**
		Ni	234.663	77			W	232.656	15
		W	234.669	200			W	232.670	19
		V	234.701	850			Ge	232.792	350
		Ni	234.751	39	341.476	4.8	Co	341.233	47
231.234	4.4	V	231.017	940			Co	341.263	26
		Ni	231.096	1.7			W	341.296	180
		Co	231.136	930			Ni	341.347	240
		V	231.146	650			W	341.353	100
		PO	*231.227*				Ni	341.394	380
		V	**231.240**	**1400**			Co	341.716	240

6. The Individual Elements

Table 6.18 (continued)

$\lambda_{element}$ / nm	Rel. sens.	Matrix name	λ / nm	Rel. sens.	$\lambda_{element}$ / nm	Rel. sens.	Matrix name	λ / nm	Rel. sens.
352.454	5.3	Co	352.157	260			Co	232.314	25
		Co	352.343	250	349.296	18	Pd	348.977	890
		Fe	352.604	870			Fe	349.057	82
		Co	352.684	22			Co	349.132	430
231.366	5.9	Co	231.136	930			Co	349.568	220
		V	231.146	650	232.195	49	Ni	232.003	1.0
		Ni	231.234	4.4			V	232.014	300
		Fe	231.310	110			Fe	232.036	74
		W	231.317	3.1			Ni	232.138	7.2
		PO	*231.354*				W	232.163	6.2
		PO	*231.369*				Co	232.314	25
		Ni	231.398	10					
		W	231.417	42	361.046	110	Fe	360.886	920
		Ge	231.420	260			Pd	360.955	13
		Ti	231.430	150			Ti	361.016	95
		V	231.562	400			Ni	361.274	570
232.138	7.2	Ni	232.003	1.0	336.155	180	Ti	335.827	19
		V	232.014	300			Ti	336.099	23
		Fe	232.036	74			Ti	336.126	13
		W	232.163	6.2			Ti	336.184	56
		Ni	232.195	49			La	336.204	9.9

6.29 Palladium (Pd)

The most sensitive analytical lines for palladium are at 244.791 nm and 247.642 nm with characteristic concentrations of $c_0 = 0.064$ mg/L and $c_0 = 0.10$ mg/L, respectively, in a very fuel-lean air/acetylene flame, whereby the former exhibits a highly non-linear calibration curve. The characteristic mass at the second line, using a transversely heated graphite tube atomizer, is about $m_0 = 22$ pg.

Palladium has about 50 absorption lines of significantly different sensitivity. However, as there is only very rarely any need to determine high concentrations of palladium, only a few selected lines are shown in Table 6.19, together with information about their spectral environment.

Table 6.19: Selected absorption lines of palladium (sorted according to their relative sensitivity) and corresponding spectral environment in order to identify species with potential risk for spectral interference, or additional elements that might be determined simultaneously

$\lambda_{element}$ / nm	Rel. sens.	Matrix name	λ / nm	Rel. sens.	$\lambda_{element}$ / nm	Rel. sens.	Matrix name	λ / nm	Rel. sens.
244.791	1.0	Ti	244.613	200			Fe	276.311	23 k
		Pb	244.618	920	324.270	6.8	Ti II	324.199	
		Fe	244.771	200			*PO*	*324.257*	
		PO	*244.786*				*PO*	*324.271*	
		W	244.839	170			*OH*	*324.282*	
		Ga	245.008	8.1			*PO*	*324.284*	
247.642	1.6	W	247.415	8.1			Ni	324.306	100
		PO	*247.625*				Ti	324.380	350
		Pb	**247.638**	**320**			La II	324.513	
		PO	*247.646*		363.470	7.2	Co	363.138	780
		V	**247.654**	**1700**			Fe	363.146	770
		PO	*247.657*				W	363.194	43
		Co	247.664	300			**Ni**	**363.494**	**18 k**
		Ni	247.688	360			Ti	363.520	88
340.458	4.0	**La**	**340.451**	**18**			Ti	363.546	1.4
		Co	340.511	25			La	363.666	15
276.309	4.8	W	276.234	7.8			Ti	363.797	170

6.30 Phosphorus (P)

Phosphorus exhibits three resonance lines at 177.435 nm, 178.223 nm and 178.705 nm in the vacuum-UV, which are not accessible to normal AA spectrometers, including HR-CS AAS.

L'vov and Khartsyzov [94] were the first to report the successful determination of phosphorus using the absorption lines at 213.547 / 213.618 nm and 214.914 nm, which emanate from metastable states. However, the sensitivity ($c_0 \approx 150$ mg/L) of the determination is relatively poor, even using the nitrous oxide / acetylene flame, and a careful selection of modifiers is necessary in order to obtain a characteristic mass of about $m_0 = 5000$ pg for the determination by GF AAS [22].

6. The Individual Elements

One of the characteristics of HR-CS AAS is that no element-specific radiation source is required, and that, in principle, absorption can be measured at any wavelength, including molecular absorption bands. Huang et al. [68] investigated the possibility of using PO molecular absorption bands in an air/acetylene flame for analytical purposes. Although molecular absorption and emission of PO and HPO have been proposed previously for the determination of phosphorus, this work was done with low-resolution equipment, making it highly susceptible to all kinds of spectral interference. As shown in Figures 7.42 – 7.49 (refer to Section 7.2.14), several of the PO bands consist of a very strong and sharp band head, followed by a weak and complex band shoulder. It is obvious that, with a high-resolution instrument, measurement should be made at the band head, the FWHM of which is not much different from that of an atomic line, and not at the whole band, in order to obtain high sensitivity and freedom from spectral interferences.

From the numerous fine-structured PO bands the authors selected five PO 'lines' at 246.40 nm, 247.62 nm, 247.78 nm, 324.62 nm and 327.04 nm, respectively, which were all band heads with a FWHM of less than 20 pm. The relative sensitivity of the five 'lines' is compiled in Table 6.20, and the band head at 246.40 nm is the most sensitive, although the difference in sensitivity is not very great. Nevertheless, as can be seen from the enlarged overview spectra in Figure 6.8, the band head at 324.62 nm should be the 'line' of choice because the spectral environment is much simpler. However, any of the other 'lines' may be chosen in the case of any spectral interference, which has been carefully investigated by the authors.

A slightly fuel-rich flame with an airflow of 14 L/min and an acetylene flow of 2 L/min resulted in the best sensitivity, and a characteristic concentration of $c_0 = 70$ mg/L phosphorus was determined. The linear dynamic range for the method was found to be more than three orders of magnitude. The LOD achieved was 1.3 mg/L for phosphorus using the PO 'line' at 324.62 nm and an illumination time of 25 s.

Table 6.20: Relative sensitivity obtained for the determination of phosphorus at five PO band heads using the CP ± 1 for measurement

Wavelength / nm	Relative sensitivity
246.40	1.00
247.62	1.59
247.78	1.35
324.62	1.56
327.04	1.75

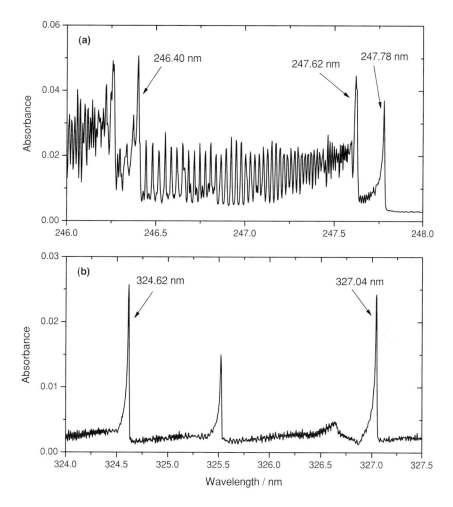

Figure 6.8: PO molecular absorbance obtained for 400 mg/L P in an air/acetylene flame at different wavelengths with band heads at (a) 246.40 nm, 247.62 nm and 247.78 nm, (b) 324.62 nm and 327.04 nm

6.31 Platinum (Pt)

The most sensitive analytical line for platinum is at 265.945 nm with a characteristic concentration of $c_0 = 0.7$ mg/L in a very fuel-lean air/acetylene flame. The characteristic mass at this line, using a transversely heated graphite tube atomizer, is $m_0 = 110$ pg. No further information is available for this element.

6.32 Potassium (K)

The most sensitive analytical line for potassium is at 766.491 nm with a characteristic concentration of $c_0 = 0.005$ mg/L in an air/acetylene flame, which is about a factor of four better than LS AAS. Potassium is significantly ionized in an air/acetylene flame, so that an excess of another easily ionized element, such as cesium, should be added in order to control this effect. The characteristic mass at this line, using a transversely heated graphite tube atomizer, is $m_0 = 1.2$ pg. A few less sensitive analytical lines that might be used for the determination of higher potassium concentrations are listed in Table 6.21, together with information about their spectral environment.

Table 6.21: Selected absorption lines of potassium (sorted according to their relative sensitivity) and corresponding spectral environment in order to identify species with potential risk for spectral interference, or additional elements that might be determined simultaneously

$\lambda_{element}$ / nm	Rel. sens.	Matrix name	λ / nm	Rel. sens.	$\lambda_{element}$ / nm	Rel. sens.	Matrix name	λ / nm	Rel. sens.
766.491	1.0				404.414	200	W	404.559	29
							K	404.721	420
769.897	1.8				404.721	420	K	404.414	200
							W	404.559	29

6.33 Rhodium (Rh)

The most sensitive analytical line for rhodium is at 343.489 nm with a characteristic concentration of $c_0 = 0.16$ mg/L in a fuel-lean air/acetylene flame. The characteristic mass at this line, using a transversely heated graphite tube atomizer, is $m_0 = 23$ pg. No further information is available about this element.

6.34 Rubidium (Rb)

The most sensitive analytical line for rubidium is at 780.027 nm with a characteristic concentration of $c_0 = 0.014$ mg/L in an air/acetylene flame, and a linear working range up to about $A_{max} = 3.3$ mg/L. Rubidium is significantly ionized in an air/acetylene flame, so that an excess of another easily ionized element, such as cesium, should be added in order to control this effect.

The characteristic mass at the main absorption line, using a transversely heated graphite tube atomizer, is $m_0 = 1.6$ pg. The sensitivity obtained with HR-CS AAS, both in a flame and in the graphite furnace, is significantly better than that with LS AAS. A few less sensitive analytical lines that might be used for the determination of higher rubidium concentrations are compiled in Table 6.22 together with information about their spectral environment.

Table 6.22: Selected absorption lines of rubidium (sorted according to their relative sensitivity) and corresponding spectral environment in order to identify species with potential risk for spectral interference, or additional elements that might be determined simultaneously

$\lambda_{element}$ / nm	Rel. sens.	Matrix name	λ / nm	Rel. sens.	$\lambda_{element}$ / nm	Rel. sens.	Matrix name	λ / nm	Rel. sens.
780.027	1.0				420.180	180			
794.760	2.0				421.553	350	Sr II	421.552	
							CN	421.553	

6.35 Ruthenium (Ru)

The most sensitive analytical line for ruthenium is at 349.895 nm with a characteristic concentration of $c_0 = 0.8$ mg/L in an air/acetylene flame. The characteristic mass at this line, using a transversely heated graphite tube atomizer, is $m_0 = 40$ pg. No further information is available about this element.

6.36 Selenium (Se)

Selenium is one of the elements most frequently determined by AAS because of its importance as an essential element for humans and many animal species, and also because the boundary between deficiency and toxic effects is extremely narrow [113, 150].

The most sensitive analytical line for selenium is at 196.026 nm with a characteristic concentration of $c_0 = 0.27$ mg/L in an air/acetylene flame. This flame already absorbs very strongly at the wavelength of selenium, and as this absorption is subject to significant fluctuation, it is recommended to apply an appropriate correction using least-squares BC. The characteristic mass obtained at the 196.026 nm line, using a transversely heated graphite tube atomizer, is about $m_0 = 40$ pg. This line is in the range of strong molecular absorption bands due to PO and NO with pronounced fine structure, which may have to be corrected for to avoid interferences. In addition there are three nearby iron lines at 195.950 nm, 196.061 nm and 196.147 nm, which, however, are well separated from the 196.026 nm selenium line using HR-CS AAS, so that no correction is necessary. The same situation arises for the absorption of the so-called Schumann-Runge bands, resulting from the oxygen existent in the optical beam path. Although these structures, which are present for wavelengths below 200 nm, can obviously be removed using least-squares BC, there is no need to do this, as they are usually not subject to any fluctuation.

As the second-most sensitive selenium line at 203.985 m, which is about a factor of four less sensitive than the primary line, exhibits even more interference from molecular absorption bands, and as selenium usually has to be determined at trace levels, no other absorption lines for selenium are discussed in this context.

6.37 Silicon (Si)

Silicon is the second most abundant element in the Earth's crust, and its omnipresence, e.g. in dust, results in an extremely high risk of contamination, particularly when traces of this element have to be determined. The most sensitive analytical line for silicon is at 251.611 nm with a characteristic concentration of $c_0 = 1.2$ mg/L in a nitrous oxide / acetylene flame. The characteristic mass at this line, using a transversely heated graphite tube atomizer, is $m_0 = 70$ pg. A number of less sensitive analytical lines that might be used for the determination of higher silicon concentrations are listed in Table 6.23, together with information about their spectral environment.

Figure 6.9 shows that a second silicon line at 251.432 nm is located within the spectral range of the primary analytical line. It is therefore possible to establish two calibration curves of significantly different slopes simultaneously, and hence to extend the working range significantly, which is of particular importance for this element.

Table 6.23: Selected absorption lines of silicon (sorted according to their relative sensitivity) and corresponding spectral environment in order to identify species with potential risk for spectral interference, or additional elements that might be determined simultaneously

$\lambda_{element}$ / nm	Rel. sens.	Matrix name	λ / nm	Rel. sens.	$\lambda_{element}$ / nm	Rel. sens.	Matrix name	λ / nm	Rel. sens.
251.611	1.0	W	251.394	68			Fe	251.625	110 k
		Si	251.432	3.5			V	251.714	53
		V	251.515	150			Co	251.786	200
		PO	*251.623*				Fe	251.810	12

6.37 Silicon (Si)

Table 6.23 (continued)

$\lambda_{element}$ / nm	Rel. sens.	Matrix name	λ / nm	Rel. sens.	$\lambda_{element}$ / nm	Rel. sens.	Matrix name	λ / nm	Rel. sens.
250.690	2.7	Co	250.452	310			V	253.018	40
		Ti	250.454	350	251.432	3.5	V	251.195	52
		W	250.470	4.8			Fe	251.236	350
		W	250.538	160			W	251.394	68
		V	250.554	890			*PO*	*251.432*	
		W	250.602	28			**V**	**251.438**	**1300**
		Cr	**250.684**	**24 k**			V	251.515	150
		Co	**250.688**	**690**			Si	251.611	1.0
		V	**250.690**	**61**	221.667	4.0	Cu	221.458	970
		V	250.778	50			W	221.480	39
		V	250.786	72			Ga	221.804	230
		W	250.874	64			Si	221.806	25
252.411	2.9	W	252.216	130			V	221.825	180
		Fe	252.285	1.7			W	221.833	48
		In	252.298	710			Ti	221.839	170
		W	252.341	9.7	221.089	5.9	Mn	220.881	67
		PO	*252.398*				W	220.906	180
		PO	*252.419*				Sn	220.965	7.6
		Ni	**252.421**	**2000**			Al	221.006	5.8
		Fe	**252.429**	**26**			Tl	221.065	140
		Ti II	252.561				**Tl**	**221.079**	**300**
		W	252.568	60			**Ni**	**221.101**	**760**
		V	252.622	16			In	221.114	620
		W	252.642	170			Ni	221.129	700
252.851	3.0	V	252.622	16			Si	221.174	17
		W	252.642	170			Ni	221.216	920
		Fe	252.744	4.8	212.412	9.2	Al	212.336	23
		Ti	252.798	63			Ti	212.355	20
		W	**252.848**	**42**			Ge	212.382	150
		PO	*252.864*				*PO*	*212.411*	
		Co	252.897	5.6			Ge	212.474	
		Fe	252.914	12			Ni	212.481	760
		Fe	252.984	74			Ni	212.563	250
		Ti	252.986	7.8			Ti	212.609	26
		Co	253.013	290					

6. The Individual Elements

Table 6.23 (continued)

$\lambda_{element}$ / nm	Rel. sens.	Matrix name	Matrix λ / nm	Rel. sens.	$\lambda_{element}$ / nm	Rel. sens.	Matrix name	Matrix λ / nm	Rel. sens.
		W	212.612	180			*NO*	*221.820*	
220.798	14	Tl	220.695	71			V	221.825	180
		Co	**220.786**	**260**			W	221.833	48
		Mn	220.881	67			Ti	221.839	170
		W	220.906	180			Ti	221.975	200
		Sn	220.965	7.6			Mo	222.025	36
		Al	221.006	5.8	205.813	67	Ti	205.675	350
288.158	15	W	287.872	35			Ti	205.713	230
		W	287.911	7.5			Ge	205.724	260
		W	287.940	4.5			*PO*	*205.809*	
		OH	*288.159*				*PO*	*205.820*	
221.806	25	Si	221.667	4.0			W II	205.830	
		Ga	**221.804**	**230**			W	205.862	42
		Pb	**221.808**	**40 k**			Ti	205.940	49
		In	**221.820**	**1600**			Ni	205.990	96

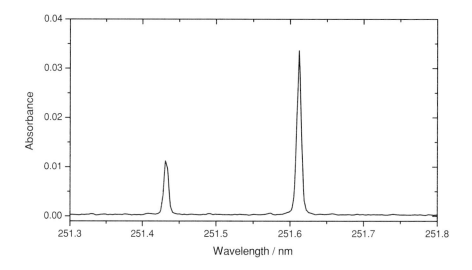

Figure 6.9: Primary silicon absorption line at 251.611 nm and a secondary line at 251.432 nm

6.38 Silver (Ag)

The most sensitive analytical line for silver is at 328.068 nm with a characteristic concentration of $c_0 = 0.02$ mg/L in an oxidizing air/acetylene flame, and a linear working range up to about $A_{max} = 1.3$ mg/L. The characteristic mass at this line, using a transversely heated graphite tube atomizer, is $m_0 = 3$ pg. A few less sensitive analytical lines that might be used for the determination of higher silver concentrations are given in Table 6.24, together with information about their spectral environment.

Table 6.24: Selected absorption lines of silver (sorted according to their relative sensitivity) and corresponding spectral environment in order to identify species with potential risk for spectral interference, or additional elements that might be determined simultaneously

$\lambda_{element}$ / nm	Rel. sens.	Matrix name	λ / nm	Rel. sens.	$\lambda_{element}$ / nm	Rel. sens.	Matrix name	λ / nm	Rel. sens.
328.068	1.0	V	327.794	780			Ti	338.594	2.6
		V	328.331	140	206.117	150	Ti	205.940	49
338.289	1.5	Ti II	338.028				Ni	205.990	96
		Ni	338.057	56			Ni	206.020	410
		Ni	338.088	410			Ti	206.059	33
		La II	338.091				*NO*	*206.106*	
		La	338.143	11			W	206.152	160
		Ti	338.231	230			Cr II	206.158	
		Co	**338.296**	**36 k**			Bi	206.163	6.2
		Mo	338.462	610			Ti	206.171	29
		Co	338.522	530			Ti	206.287	33
		Ti	338.566	25			W	206.311	64

6.39 Sodium (Na)

The most sensitive analytical line for sodium is at 588.995 nm with a characteristic concentration of $c_0 = 0.0024$ mg/L in an oxidizing air/acetylene flame, and a linear working range up to about $A_{max} = 0.35$ mg/L. Sodium is significantly ionized in an air/acetylene flame, so that an excess of another easily ionized element, such as cesium, should be added in order to control this effect. The characteristic mass at the 588.995 nm line, using a transversely heated graphite tube atomizer, is $m_0 = 0.2$ pg. The sensitivity obtained with HR-CS AAS, both in a flame and in the graphite furnace, is significantly better than that with LS AAS.

6. The Individual Elements

Several less sensitive analytical lines that might be used for the determination of higher sodium concentrations are listed in Table 6.25, together with information about their spectral environment.

All sodium lines are doublets, which are well resolved. As shown in Figure 6.10 for the doublet at 330.237 / 330.298 nm, both doublet lines are within the spectral window evaluated by the detector, so that both lines may be used simultaneously for analytical purposes.

Table 6.25: Selected absorption lines of sodium (sorted according to their relative sensitivity) and corresponding spectral environment in order to identify species with potential risk for spectral interference, or additional elements that might be determined simultaneously

$\lambda_{element}$ / nm	Rel. sens.	Matrix name	λ / nm	Rel. sens.	$\lambda_{element}$ / nm	Rel. sens.	Matrix name	λ / nm	Rel. sens.
588.995	1.0						Pd	330.213	150
589.592	2.0						Na	330.298	370
330.237	210	Ti	329.941	330	330.298	370	W	330.082	25
		W	330.082	25			Pd	330.213	150
							Na	330.237	210

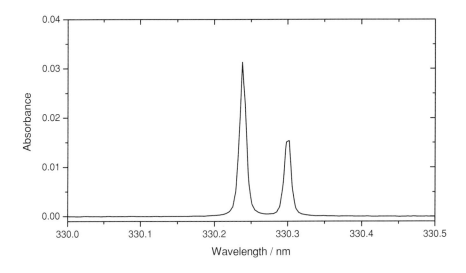

Figure 6.10: Secondary sodium absorption line doublet at 330.237 / 330.298 nm

6.40 Strontium (Sr)

The most sensitive analytical line for strontium is at 460.733 nm with a characteristic concentration of $c_0 = 0.08$ mg/L in a nitrous oxide / acetylene flame. The characteristic mass at this line, using a transversely heated graphite tube atomizer, is $m_0 = 1.2$ pg. A few less sensitive analytical lines that might be used for the determination of higher strontium concentrations are listed in Table 6.26, together with information about their spectral environment.

Table 6.26: Selected absorption lines of strontium (sorted according to their relative sensitivity) and corresponding spectral environment in order to identify species with potential risk for spectral interference, or additional elements that might be determined simultaneously

$\lambda_{element}$ / nm	Rel. sens.	Matrix name	λ / nm	Rel. sens.	$\lambda_{element}$ / nm	Rel. sens.	Matrix name	λ / nm	Rel. sens.
460.733	1.0						Co	235.548	140
242.810	62	V	242.613	940	230.730	130	Co	230.517	190
		V	242.774	670			Ti	230.569	5.2
		PO	*242.815*				W	230.659	17
		PO	*242.822*				**Ni**	**230.736**	**320**
		Ti	242.822	77			Ti	230.890	130
		V	242.828	58			Fe	230.900	180
		Ti	242.838	160			Co	230.901	14
		Co	242.923	380			W	230.902	22
		Sn	242.949	3.3	256.947	170	Co	256.734	63
		W	242.985	70			W	256.750	72
		In	242.986	870			Al	256.798	9.3
235.430	69	Co	235.285	170			W	256.821	53
		Co	235.337	92			W	256.856	55
		Cr	**235.431**	**16 k**			**Fe**	**256.960**	**43 k**
		W	235.461	14			W	257.009	94
		Sn	235.484	1.4			Sn	257.158	590
		Ni	235.505	260					

6.41 Sulfur (S)

Sulfur cannot be determined directly by AAS since its main resonance line at 180.671 nm, as well as two other lines at 181.974 nm and 182.565 nm are in the vacuum-UV, and hence

6. The Individual Elements

not accessible with conventional instrumentation. A number of indirect methods have been described for the determination of sulfur by AAS [150], however, these methods have never found general acceptance, as they are all rather laborious and time consuming.

Huang et al. [70] investigated the possibility of determining sulfur with HR-CS AAS, using the fine structure of molecular absorption bands. As shown in Figures 7.21 – 7.23 (refer to Section 7.2.7), sulfur exhibits a relatively strong molecular absorption spectrum due to the CS radical, the most intense part of which in the wavelength range around 258 nm is shown in Figure 6.11. A very fuel-rich flame with an air flow-rate of 17 L/min and an acetylene flow-rate of 3.8 L/min was necessary to obtain the strongest CS absorption.

In principle, all the absorption lines in Figure 6.11 could be used for the determination of sulfur, as most of them have a FWHM around 5 pm, i.e. are of the same magnitude as atomic lines. The sensitivity, expressed as characteristic concentration c_0, obtained at a number of selected lines is shown in Table 6.27, which indicates that the line at 257.594 nm exhibits the best sensitivity, but that the sensitivity at all the other lines is within less than a factor of two.

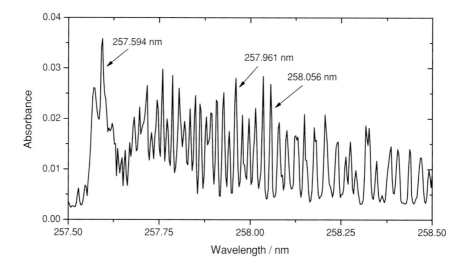

Figure 6.11: CS molecular absorption of 1% v/v sulfuric acid solution recorded by the high-resolution simultaneous absorption spectrometer (ARES) using a fuel-rich air/acetylene flame

Two of the lines were investigated in more detail, the line at 257.594 nm, because it exhibited the best sensitivity, and the line at 258.056 nm, because it could be very well iso-

Table 6.27: Characteristic concentration (c_0) obtained for the determination of sulfur at selected CS molecular lines in a fuel-rich air/acetylene flame

Wavelength / nm	Char. concentration / mg L^{-1}
257.594	90
257.961	120
258.056	125
258.035	131
257.570	137
257.927	139
257.907	148

lated from other absorption lines. Firstly, the influence of various acids was investigated, and it was found that HNO$_3$, HCl and HF up to a concentration of 5 % v/v and HClO$_4$ up to a concentration of 2 % v/v had no influence on the absorbance signal. Secondly, the influence of 12 common matrix elements up to a concentration of 2 g/L on the absorbance of 400 mg/L sulfur was investigated. None of the elements, including iron, caused any spectral interference in the determination of sulfur at these lines. Only lead at concentrations above 0.2 mg/L caused a non-spectral interference due to the precipitation of lead sulfate.

The calibration curves for both lines were strictly linear up to at least 3000 mg/L sulfur, and deviate from linearity only slightly up to 6000 mg/L sulfur, i.e. the working range covers more than three orders of magnitude. The best LOD of 2.5 mg/L was obtained at the 258.056 nm line, compared to about 5 mg/L at the 257.594 nm line, so that preference should be given to the former one.

6.42 Tellurium (Te)

The most sensitive analytical line for tellurium is at 214.281 nm with a characteristic concentration of $c_0 = 0.18$ mg/L in an air/acetylene flame. The characteristic mass at this line, using a transversely heated graphite tube atomizer, is $m_0 = 17$ pg. The analytical line at 214.281 nm, as well as that at 225.902 nm, which is about a factor of four less sensitive, is within the range of strong NO absorption bands, which requires special attention in the case of a matrix that contains nitrogen compounds at high concentration.

6.43 Thallium (Tl)

The most sensitive analytical line for thallium is at 276.787 nm with a characteristic concentration of $c_0 = 0.13$ mg/L in an air/acetylene flame. The characteristic mass at this line, using a transversely heated graphite tube atomizer, is $m_0 = 10$ pg. The sensitivity for thallium obtained with HR-CS AAS, both in a flame and in the graphite furnace, is more than a factor of two better than that with LS AAS. The most sensitive analytical line of thallium is within the range of strong S_2 absorption bands, which requires special attention in the case of a high sulfur-containing matrix. The spectral interference due to palladium, which makes the use of this modifier impossible in LS AAS, is absent in HR-CS AAS, as the palladium line at 276.309 nm is well separated from the thallium line. A number of less sensitive analytical lines that might be used for the determination of higher thallium concentrations are given in Table 6.28, together with information about their spectral environment.

Table 6.28: Selected absorption lines of thallium (sorted according to their relative sensitivity) and corresponding spectral environment in order to identify species with potential risk for spectral interference, or additional elements that might be determined simultaneously

$\lambda_{element}$ / nm	Rel. sens.	Matrix name	λ / nm	Rel. sens.	$\lambda_{element}$ / nm	Rel. sens.	Matrix name	λ / nm	Rel. sens.
276.787	1.0	OH	276.780				W II	223.706	
		W	276.898	14			W	223.721	83
		W	276.974	45			V	223.723	170
377.572	1.8	Ni	377.557	1300			W	223.769	70
		V	377.868	220			Sr	223.770	1200
							NO	223.779	
237.958	4.4	Ti	237.815	76			Ti	223.822	100
		Al	237.839	64			Ti	223.875	17
		Ge	237.914	470	216.855	44	Fe	216.677	3.5
		PO	237.942				W	216.769	140
		PO	237.966				W	216.830	46
		Ni	237.972	6700			NO	216.854	
		Co	238.048	46			Co	216.871	520
		Sn	238.072	130			Al	216.883	18
		Ti	238.081	130			W	216.948	6.2
		Fe	238.184	720			Al	216.984	530
223.780	13	Ga	223.590	400			Pb	217.001	1.0
		Co	223.680	160			W	217.020	87

6.44 Tin (Sn)

The most sensitive analytical line for tin is at 224.605 nm with a characteristic concentration of $c_0 = 1.3$ mg/L in an air/acetylene flame. However, other flames, such as an air/hydrogen or a nitrous oxide/acetylene flame may be used as well, depending mostly on the matrix. The characteristic mass at this line, using a transversely heated graphite tube atomizer, is $m_0 = 40$ pg. A number of less sensitive analytical lines that might be used for the determination of higher tin concentrations are listed in Table 6.29, together with information about their spectral environment.

Table 6.29: Selected absorption lines of tin (sorted according to their relative sensitivity) and corresponding spectral environment in order to identify species with potential risk for spectral interference, or additional elements that might be determined simultaneously

$\lambda_{element}$ / nm	Rel. sens.	Matrix name	λ / nm	Rel. sens.	$\lambda_{element}$ / nm	Rel. sens.	Matrix name	λ / nm	Rel. sens.
224.605	1.0	Cu	224.426	260			V II	270.617	
		Ni	224.448	110			*OH*	*270.633*	
		Ni	224.453	320			*OH*	*270.649*	
		Ti	224.469	98			Fe	270.658	3400
		V	224.576	490			W	270.880	66
		NO	*224.597*				W	270.893	39
		Co	224.660	810	242.949	3.3	V	242.774	670
		W	224.677	11			Sr	242.810	62
		Pb	224.686	360			Ti	242.822	77
235.484	1.4	Co	235.285	170			V	242.828	58
		Co	235.337	92			Ti	242.838	160
		Sr	235.430	69			Co	242.923	380
		W	235.461	14			*PO*	*242.940*	
		NO	*235.481*				Na	242.940	2900 k
		NO	*235.498*				W	242.985	70
		Ni	235.505	260			In	242.986	870
		Co	235.548	140			W	243.108	8.5
		Ni	235.687	190			V	243.157	750
286.332	1.6	W	286.301	87	283.999	4.3	W	283.734	66
		V	286.436	35			W	284.157	21
		W	286.606	8.4	303.412	4.5	Ni	303.187	340
270.651	2.9	W	270.601	140			Cr	*303.419*	1700

Table 6.29 (continued)

$\lambda_{element}$ / nm	Rel. sens.	Matrix name	λ / nm	Rel. sens.	$\lambda_{element}$ / nm	Rel. sens.	Matrix name	λ / nm	Rel. sens.
		Cr	303.704	910			**Co**	**226.875**	**4000**
219.934	5.2	Ni	219.735	94			**In**	**226.889**	**7300**
		In	219.741	680			Al	226.910	3.6
		W	219.837	66			W II	227.015	
		Ge	219.871	13	219.449	20	W	219.303	18
		W	219.892	11			V	219.466	720
		Al	219.918	570			Ga	219.533	520
		W	**219.930**	**110**			Fe	219.604	26
		Cu	219.958	390			V	219.641	200
		Ga	219.963	830			Co	219.646	64
		V	220.018	460	228.668	35	W	228.480	21
		Fe	220.039	61			W	228.491	75
		Ni	220.068	580			W	228.517	18
		Fe	220.072	83			**PO**	**228.664**	
		Ca	220.073	140			Fe	228.725	110
		W	220.151	44			Ni	228.733	370
254.655	5.3	Co	254.425	31			Co	228.781	110
		W	254.534	11			Cd	228.802	1.0
		V	254.592	81			Ni	228.840	240
		Fe	254.598	18	242.170	79	Ni	241.931	120
		PO	**254.637**				V	242.012	71
		PO	**254.657**				V	242.106	75
		Co	**254.662**	**18 k**			Ni	242.123	680
		PO	**254.667**				Ti	242.130	20
		W	254.714	4.5			**Co**	**242.169**	**39 k**
		Mo	254.822	810			V	242.198	64
226.891	5.8	Fe	226.709	320			W	242.229	36
		Sn	226.719	960			V	242.337	280
		Ti	226.737	150			Ni	242.403	750
		Ti	226.754	13	202.698	170	Ni	202.539	230
		Ni	226.756	750			Mg	202.582	23
		Ti	226.791	45			Ni	202.659	73
		Co	226.818	940			**NO**	**202.706**	
		W	226.865	80			V	202.764	200
		Ti	**226.875**	**56**			V	202.841	200

6.45 Titanium (Ti)

The most sensitive analytical line for titanium in HR-CS AAS is at 319.200 nm with a characteristic concentration of about $c_0 = 1$ mg/L in a nitrous oxide/acetylene flame. The characteristic mass at this line, using a transversely heated graphite tube atomizer, is $m_0 = 45$ pg.

Although titanium has more than 300 absorption lines, only a few of the most sensitive ones are listed in Table 6.30, together with information about their spectral environment, because the sensitivity of titanium is not very high anyway, so that there is essentially no need to reduce the sensitivity further. The only aspect of interest might therefore be the simultaneous determination of other analytes together with titanium.

Table 6.30: Selected absorption lines of titanium (sorted according to their relative sensitivity) and corresponding spectral environment in order to identify species with potential risk for spectral interference, or additional elements that might be determined simultaneously

$\lambda_{element}$ / nm	Rel. sens.	Matrix name	λ / nm	Rel. sens.	$\lambda_{element}$ / nm	Rel. sens.	Matrix name	λ / nm	Rel. sens.
319.200	1.0	W	319.157	18			Co	335.438	520
		OH	*319.200*		364.267	1.3	Pb	363.957	290
		Mo	319.397	1.8			La	364.153	2.4
319.992	1.1	V	319.801	18			La II	364.542	
		W	319.884	57			Ti	364.620	60
		V	320.239	19	365.350	1.3	Ti	365.459	27
335.463	1.3	Ti	335.293	85					

6.46 Tungsten (W)

Tungsten can be determined with a nitrous oxide/acetylene flame at 255.135 nm with a characteristic concentration of $c_0 = 6$ mg/L. Due to its extremely high melting point and its tendency to form stable carbides, tungsten cannot be determined by GF AAS. Although tungsten has more than 300 absorption lines, only a few of the most sensitive ones are listed in Table 6.31, together with information about their spectral environment, because the sensitivity of tungsten is not very high anyway, so that there is essentially no need to reduce the sensitivity further. The only aspect of interest might therefore be the simultaneous determination of other analytes together with tungsten.

Table 6.31: Selected absorption lines of tungsten (sorted according to their relative sensitivity) and corresponding spectral environment in order to identify species with potential risk for spectral interference, or additional elements that might be determined simultaneously

$\lambda_{element}$ / nm	Rel. sens.	Matrix name	λ / nm	Rel. sens.	$\lambda_{element}$ / nm	Rel. sens.	Matrix name	λ / nm	Rel. sens.
255.135	1.0	Co	254.929	490			W	272.503	18
		Fe	254.961	25			Ti	272.507	180
		V	254.997	390			La	272.557	5.6
		W	255.038	7.3			Cr	272.650	710
		W	255.100	43	240.558	2.3	V	**240.548**	**1600**
		PO	*255.123*				W	**240.569**	**9.9**
		PO	*255.134*				W	240.618	23
		PO	*255.149*				V	240.675	85
		V	255.265	67			Co	240.725	1.0
		Co	255.300	960			V	240.790	140
		W	255.316	120					
		Co	255.337	890	243.596	2.3	PO	*234.587*	
		W	255.382	22			PO	*234.603*	
245.200	1.9	Ga	245.008	8.1			W	243.398	10
		W	245.135	83			Ti	243.410	140
		W II	245.148				V	243.552	57
		PO	*245.194*				W	**243.579**	**68**
		PO	*245.213*				Co	**243.582**	**480**
		Ni	245.399	890			W	243.626	80
							W	243.662	17
272.435	2.1	V	272.256	520			Co	243.666	8.9
		Fe	272.358	30			Ti	243.829	330
		W	272.462	77					

6.47 Vanadium (V)

Vanadium has an absorption line triplet at 318.341 / 318.397 / 318.538 nm, as shown in Figure 6.12. The most sensitive analytical line is at 318.397 nm with a characteristic concentration of $c_0 = 0.28$ mg/L in a nitrous oxide / acetylene flame, but all the other lines of the triplet may be used as well. The characteristic mass at this line, using a transversely heated graphite tube atomizer, is $m_0 = 20$ pg.

Vanadium also has more than 300 absorption lines, but only a selection of the most sensitive ones are compiled in Table 6.32, together with information about their spectral environment. These lines may be used for the determination of higher vanadium concentrations, or for the simultaneous determination of other analytes together with vanadium.

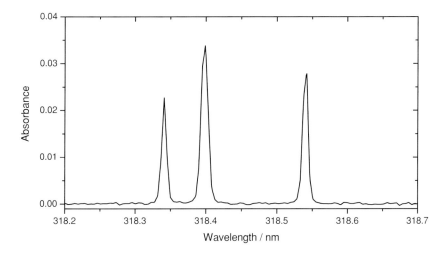

Figure 6.12: Vanadium absorption line triplet at 318.341 / 318.397 / 318.538 nm

Table 6.32: Selected absorption lines of vanadium (sorted according to their relative sensitivity) and corresponding spectral environment in order to identify species with potential risk for spectr. interference, or additional elements that might be determined simultaneously

$\lambda_{element}$ / nm	Rel. sens.	Matrix name	λ / nm	Rel. sens.	$\lambda_{element}$ / nm	Rel. sens.	Matrix name	λ / nm	Rel. sens.
318.397	1.0	V	318.341	1.5			W	318.442	180
		OH	*318.391*				V	318.538	1.2
		W	318.442	180			Ti	318.645	1.6
		V	318.538	1.2	306.637	2.6	V	306.372	450
		Ti	318.645	1.6			Ni	306.462	120
318.538	1.2	V	318.341	1.5			Pd	306.531	210
		V	318.397	1.0			*Ti II*	*306.623*	
		W	318.442	180			*OH*	*306.631*	
		Ti	318.645	1.6			*Ti II*	*306.635*	
318.341	1.5	*OH*	*318.350*				*V*	*306.652*	*29*
		V	318.397	1.0			Bi	306.772	4.2

Table 6.32 (continued)

$\lambda_{element}$ / nm	Rel. sens.	Matrix name	λ / nm	Rel. sens.	$\lambda_{element}$ / nm	Rel. sens.	Matrix name	λ / nm	Rel. sens.
306.046	3.5	Fe	305.745	590			Pd	209.263	810
		Ni	305.764	30			Mn	209.340	90
		Fe	305.909	25			Fe	209.368	66
		Co	306.182	41			W	209.372	80
305.633	4.9	V	305.365	8.3			Al	209.407	130
		Cr	305.387	640			W	209.408	120
		Ni	305.431	36			Ge	209.426	1.3
		Fe	305.745	590			Sn	209.435	94
		Ni	305.764	30	385.584	8.2	V	385.536	23
		Fe	305.909	25			**CN**	**385.563**	
209.244	7.2	Fe	209.038	600			Fe	385.637	110
		W	209.048	5.9			Ni	385.830	840
		V	209.054	360	370.471	20	V	370.357	12
		V	209.069	21			**La**	**370.454**	**7.4**
		V	209.090	120			V	370.503	54
		V	209.131	28			Fe	370.557	130
		Sn	209.158	11			W	370.792	29
		Al	209.169	190	368.807	49	Ti	368.734	400
		Al	209.191	170			Ti	368.992	30
		V	209.191	590			V	369.029	65
		Mn	209.216	69			Pd	369.034	600
		V	**209.234**	**23**	329.815	190	Ti	329.941	330
		Mn	**209.250**	**180**			W	330.082	25
		W	**209.254**	**48**					

6.48 Zinc (Zn)

Zinc makes up about 0.012 % of the Earth's crust, i.e. it is quite abundant, which results in significant contamination problems, particularly because it is the most sensitive element in AAS. Zinc has essentially only one usable wavelength for its determination at 213.857 nm with a characteristic concentration of $c_0 = 0.007$ mg/L in an air/acetylene flame. The characteristic mass at this line, using a transversely heated graphite tube atomizer, is $m_0 = 0.5$ pg.

There is a direct line overlap between an iron line at 213.859 nm with the zinc line, which has to be taken into consideration whenever zinc has to be determined in the presence of high iron concentrations. However, the presence of a second iron line within the spectral window of HR-CS AAS at 213.970 nm makes a correction of this interference easy using least-squares BC (refer to Section 8.1.4). There is also a direct line overlap of the zinc resonance line with a weak copper line at 213.853 nm, which requires special attention in the presence of high copper containing matrices (refer to Section 8.1.5).

The only other absorption line of zinc at 307.590 nm is of little analytical significance, as the sensitivity at this line is about a factor of 5000 lower than at the primary resonance line.

7. Electron Excitation Spectra of Diatomic Molecules

7.1 General Considerations

As already mentioned in Section 2.3, molecules show complex and fine-structured spectra. This is due to the numerous rotational, vibrational and electronic transitions possible between the different energy levels. While the first two types of transitions are present in the microwave to IR region, the electronic ones mostly lie in the visible to UV wavelength range and therefore are the dominant causes of spectral interference with atomic lines or erroneous background correction (BC).

Molecules or radicals are present all the time during sample analysis, not necessarily coming only from the sample matrix but also from the atomizer itself, e.g. the flame gases. Because of its spectrally resolved signal registration scheme HR-CS AAS offers the unique possibility to correct for this type of background via mathematical procedures based on a least-squares fitting algorithm (for details see Section 5.2.2). However, this presumes knowledge and availability of the corresponding reference spectra pre-recorded with high spectral resolution as well as good accuracy in wavelength. Thinking of all possible combinations of the species present in the atomizer, the total number of molecular spectra that have to be taken into account would be incalculable. Fortunately only diatomic molecules produce spectra with fine structures having comparable widths to those of the atomic lines [40]. In contrast, polyatomic molecules show broad or smooth spectra on the picometer scale because of the strongly enhanced and non-resolvable line density. They are of minor importance and can be handled as 'continuous' background during signal processing (refer to Section 5.2.1). Therefore the huge number of molecules relevant for BC is reduced dramatically, and the situation becomes better manageable, but still requires a data base of diatomic molecular spectra covering at least the most notorious troublemakers.

7. Electron Excitation Spectra of Diatomic Molecules

Over the years many publications have focused on this problem, nevertheless most of the investigations suffer from either too low spectral resolution [24, 25, 75, 76, 111] or limited spectral range [7,52]. This is understandable, since the recording of these overview spectra requires special spectroscopic equipment, capable of simultaneous registration of the wavelength range relevant to the majority of the AAS elements (about 200 nm to 465 nm), in combination with a high spectral resolution comparable to the width of atomic lines. Such equipment was not available in the past. Furthermore, assuming a conventional system with a stepwise scanning monochromator, e.g. a wavelength increment of 5 pm with a time cycle of one second per measurement, the registration time for a total spectrum with 50 000 individual data points would lie in the range of about 14 h. Additionally taking into account that sample and blank spectra have to be recorded under identical conditions, the scanning measurement proves impossible.

The progress in solid-state image detector devices gave new impulse to the field of simultaneous spectrum recording. Nowadays compact high-resolution echelle-based spectrographs with CCD registration are feasible [9]. The combination of a prism and a grating allows a two-dimensional arrangement of the different echelle orders to be recorded easily by an array detector. A typical example of such an image is shown in Figure 7.1, where more than 100 diffraction orders have been simultaneously recorded by a 1024×1024 pix-

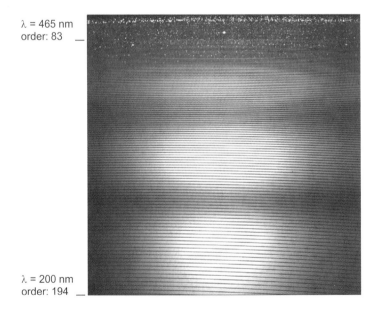

Figure 7.1: Two-dimensional echelle spectrum of a deuterium (D_2) lamp recorded with a 1024×1024 pixels CCD image detector (original size 25×25 mm^2)

els image detector (Kodak KAF 1000 with UV coating). The detailed description of the instrument, named ARES, can be found in Section 3.2.3. Since the intensity distribution of the deuterium (D_2) lamp is not linear outlined and overlaid by the blaze efficiency of the echelle grating in each order, further image processing must be applied to transform the pixel-based two-dimensional echelle spectrum into a common one-dimensional intensity over wavelength plot. Initially, this process vertically bins the pixel intensities across the height of an individual order, corresponding to a certain wavelength, then merges these values in order to obtain a contiguous spectrum and finally performs a linear interpolation resulting in equidistant wavelength increments.

In this way, intensity spectra with high spectral resolution, covering the wavelength range of major analytical interest, can be measured within a few seconds per individual illumination, as shown in Figure 7.2 (a) and (b). All order segments are now merged together but the progression of the blaze efficiency of the echelle grating is still visible. In Figure 7.2 (a) the pure spectrum of the D_2 lamp without the flame ignited is depicted, whereas Figure 7.2 (b) shows the intensity distribution with the air/acetylene flame turned on. The resulting transmittance spectrum, calculated from the quotient of both spectra, is displayed in Figure 7.2 (c). It includes strong OH flame-absorption bands, representing low-energy vibrations within the electronic transition starting from the $X\,^2\Pi$ ground to the $A\,^2\Sigma^+$ state. For further details refer to Section 7.2.13.

In total, a dispersion of about $\lambda/80\,000$ per pixel is obtained with the ARES spectrograph. This means, of course, that in many cases the line widths of the electronic transitions of molecules in flames are still narrower than the full width at half maximum (FWHM) of the instrument function, which is sampled only by about two pixels. As a result, a fundamental problem in calculation of the analytical signal arises. Assuming a small wavelength drift between the recording of the reference and sample spectrum, the quotient will generate spurious absorbance values at the wings of narrow lines, even if the absolute signal intensity remains constant. A mathematical correction of such artifacts is not possible for a pixel-dominated line profile, since, according to the sampling theorem, at least seven intensity values are required over the entire profile for its full reconstruction.

One possibility to overcome the problem could be the application of an even higher spectral resolution. This would imply the use of larger optical components to improve the linear dispersion and an enlarged CCD array detector, increasing the read-out time per spectrum and the total costs of the spectrograph dramatically. The better and more elegant way is the implementation of an active wavelength stabilization, as is done in the ARES spectrograph (see Section 3.2.3). Using micro-positioning units, based on piezo-electric actuators, the spectrum is actively stabilized regarding its position and dispersion with respect to the pixel matrix of the CCD array detector. As reference, the simultaneously recorded emission lines of an internal neon lamp are used.

7. Electron Excitation Spectra of Diatomic Molecules

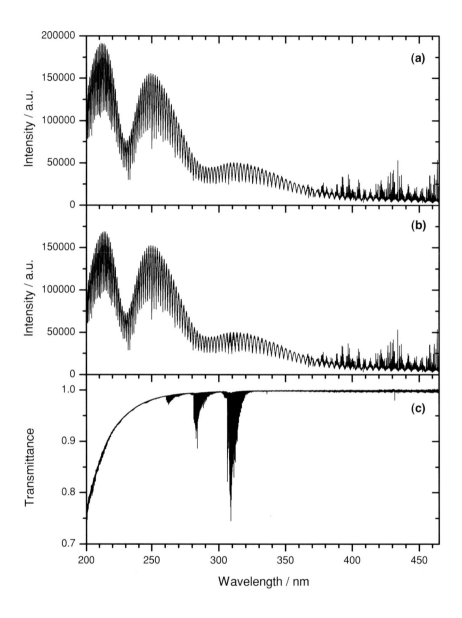

Figure 7.2: Overview spectra recorded with the ARES spectrograph; (a) deuterium lamp without flame; (b) deuterium lamp with air / acetylene flame; (c) resulting calculated flame transmittance

7.1 General Considerations

After each illumination the barycenters of intensity are calculated and corrected for to give an absolute position stability of better than 1 % (rms, i.e 1σ) of the pixel width for the long wavelength reference lines. Even in the case of temperature changes of more than 10 °C the maximum displacements of any sample line from its target position are less than 20 % of the pixel width. This result was obtained from the long time measurement over 87 h shown in Figure 7.3. In addition, the active stabilization not only corrects for thermal effects but also for day-to-day changes such as variations of the index of refraction of the air and micro-mechanical drifts. For a typical absorbance measurement over about 0.5 h, and with a temperature drift of less than 1 K/h, the stabilization of all line positions is better than 1 % (rms) of the pixel width.

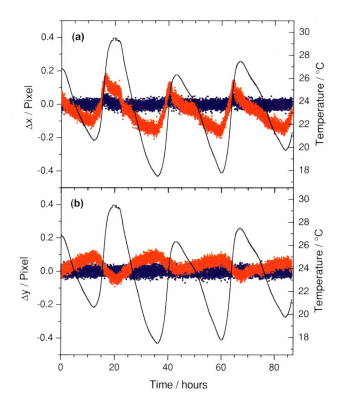

Figure 7.3: Efficiency of the active spectrum stabilization; (a) displacements parallel to the echelle dispersion; (b) displacements perpendicular to the echelle dispersion; (black: spectrograph temperature; blue: position of the Ne reference line at 614.306 nm; red: position of the Zn sample line at 213.856 nm)

7. Electron Excitation Spectra of Diatomic Molecules

This positioning accuracy is good enough for the intended application, as can be seen from the numerical simulation shown in Figure 7.4. Diagram (a) shows a calculated absorption line (Voigt profile) with 90 % center transmittance and the expected response of the CCD array whose pixel distance equals half the line width, whereas in diagram (b) the Voigt profile is slightly displaced to the right by 10 % of the pixel width, which corresponds to a shift of $\lambda/800\,000$. The ratio of the pixel responses of the two profiles is depicted in diagram (c). This simulation demonstrates that a position shift in the region of 10 % of the pixel width causes spurious absorption values of about 0.6 % on the transmittance signal. Hence, the realized stabilization of better than 1 % (rms) of the pixel width, corresponding to a wavelength stabilization of about $\lambda/8\,000\,000$ obtainable with the ARES, is sufficient for the detection of absorptions at a level of 0.1 % (sub-milli absorbance). This is even valid in a spectral region with strongly structured flame absorptions.

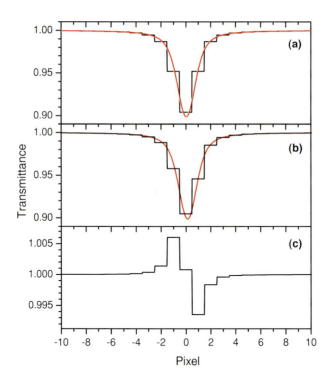

Figure 7.4: Influence of a line shift on the transmittance signal; (a) calculated Voigt profile (red line) and corresponding pixel response (black line); (b) same profile shifted 10 % of the pixel width to the right; (c) resulting pixel-based quotient

7.2 Individual Overview Spectra

All the following spectra of individual molecules were measured with the ARES spectrograph described in Section 3.2.3. The complete experimental setup for the overview absorbance measurements is shown in Figure 7.5. The atomizer used was a standard flame module of an AAnalyst 800 spectrometer (Bodenseewerk Perkin-Elmer, Überlingen, Germany) and a high-power deuterium lamp (D200 F, Heraeus, Hanau, Germany) as well as a tungsten halogen lamp (12 V, 100 W, Osram, Berlin, Germany) were applied as continuum sources (CS).

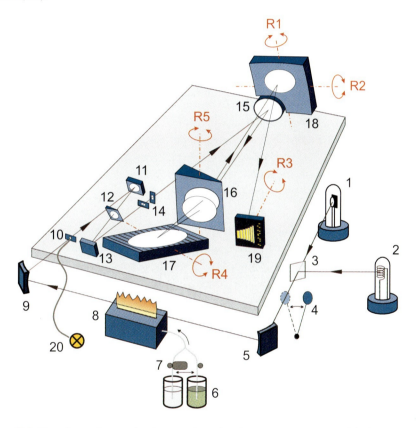

Figure 7.5: Experimental setup for the overview absorbance measurements with the ARES spectrograph; (1) D_2 lamp, (2) tungsten halogen lamp, (3) beam combiner, (4) mechanical shutter, (5,9) toroidal mirrors, (6) sample and blank reservoir, (7) magnetic valve, (8) flame atomizer, (10) entrance slit, (11 – 14) pre-dispersing illumination optics, (15,18) spherical mirrors, (16) prism, (17) echelle grating, (19) CCD array detector, (20) Ne lamp, (R1 – R5) piezo-electrically controlled rotation units

7. Electron Excitation Spectra of Diatomic Molecules

In most cases the spectra were recorded using a conventional air/acetylene flame. The measuring cycle consisted of an initial registration of the dark signal, followed by a series of alternating recordings of blank and sample intensity distributions. The solutions were changed automatically by means of a computer-controlled magnetic valve. Both groups of spectra were averaged and subsequently divided to give transmittance values. Finally, their logarithm was calculated to obtain the overview spectrum in absorbance units. The individual recording time (including data transfer) for each image was of the order of 20 s resulting in a total time of about 30 min for the whole registration cycle, usually 50 blank and 50 sample measurements. Using this method high-resolution (about 67 000 spectral channels representing the entire wavelength range from 200 nm to 465 nm) overview absorbance spectra could be recorded simultaneously. Moreover, due to the multitude of averaged spectra the transmittance noise was kept at less than 0.03 % (rms) for wavelengths below 350 nm. The noise increased to longer wavelengths because of the D_2 lamp characteristics, but was still acceptable in most cases.

When necessary a tungsten halogen lamp was additional used to enhance the intensity for wavelengths above 350 nm. Both lamp radiations were overlaid using a beam combiner and simultaneously passed through the flame. In this way, even at the long wavelength end, low noise levels were obtained. However, the overall radiation intensity in the UV was reduced as a result of the introduction of the beam combiner, resulting in longer exposure times, if comparable noise levels are needed.

Especially in the case of measurements with the nitrous oxide/acetylene flame, a further modification was necessary, since emission effects became more severe and had to be corrected for. Here the registration scheme was changed as follows: First, a blank spectrum was recorded with the CS turned on. Then, the beam pass was blocked by a shutter and a second blank spectrum was recorded. Subtraction of the latter spectrum, which represents the emissions coming from the blank and flame, gives the emission-corrected blank intensity spectrum. The same procedure was carried out for the sample solution. Again, series of this cycle were recorded to improve the SNR. Last but not least, the two groups of emission-corrected spectra were averaged, divided, and their logarithm gave the absorbance spectrum. This procedure, of course, had the highest sample consumption due to its doubled illumination time, but on the other hand provided the best SNR for longer wavelengths as well as the cleanest absorbance spectra.

In the next sections the reader will find a compilation of diatomic molecules producing fine-structured background absorptions, but it should be pointed out that this list does not claim completeness. The individual spectra are shown in total over the respective wavelength range as well as in several enlarged clippings with interesting structures and band heads. Even then it is impossible to reflect all the information included in the original data sets without seriously increasing the size of this book. Nevertheless, a lot of additional information is given for band heads, band origins, and/or rotational lines of selected band systems starting at the electronic ground state (0 eV) to enable an easy identification of these spectra when compared with others obtained with different instrumentation. Data about rotational transitions are not presented, because they are unavailable for the molecules of interest.

Spectral data are taken from the *PLASUS SpecLine* software library for atoms, ions and molecules as well as directly from the corresponding literature [108, 122]. *PLASUS Specline* is a commercially available program, but all the information is based exclusively on public sources and data banks [20, 84, 115, 139, 158]. For further details about *PLASUS SpecLine* contact PLASUS Ingenieurbüro, Dr.-Ing. Thomas Schütte, Robert-Koch-Straße 8, 86343 Königsbrunn, Germany or visit the web on http://www.plasus.de.

The relative intensities found in the literature and listed in the subsequent tables were mainly measured in emission and should be taken more as approximate values since they are not directly comparable with those obtained in absorption. Nevertheless they are valid for getting a quick overview of the most intense transitions.

7.2.1 AgH

The absorbance spectrum of the AgH molecule was measured in an air/acetylene flame with the standard registration cycle, consisting of 2×50 individual measurements with an illumination time of 17 s each. A solution of 10 g/L Ag as $AgNO_3$ dissolved in 1 % v/v HNO_3 was used as the sample and deionized water as the reference. The overview spectrum in Figure 7.6 shows four strong Ag lines at 206.117 nm, 206.985 nm, 328.068 nm and 338.289 nm, as well as AgH molecular structures around 336 nm resulting from a $\Delta v = 0$ sequence (i.e. no change of the vibrational excitation) of the $X^1\Sigma^+ \to A^1\Sigma^+$ electronic transition. The corresponding characteristics of the bands are listed in Table 7.1. In addition, another structure is visible around 217 nm which is not listed in the literature. Most probably this structure is due to a diatomic molecule containing Ag, such as AgH. Spectrum details are depicted in Figures 7.7 and 7.8.

Table 7.1: Characteristics of the AgH bands

Wavelength / nm	Relative intensity	Upper electronic energy / eV	Electronic transition	Vibrational transition
333.00	?	3.71	$X^1\Sigma^+ \to A^1\Sigma^+$	$0 \to 0$
335.70	?	3.71	$X^1\Sigma^+ \to A^1\Sigma^+$	$1 \to 1$

7. Electron Excitation Spectra of Diatomic Molecules

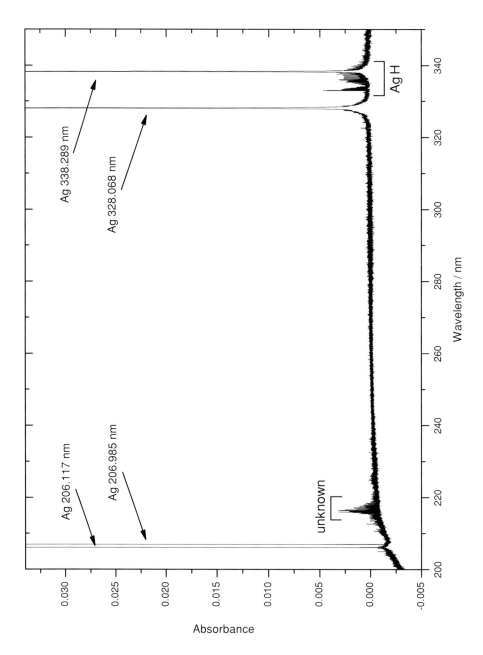

Figure 7.6: Overview spectrum of Ag in an air/acetylene flame with bands of AgH and an unknown molecule (sample: 10 g/L Ag as AgNO$_3$ in 1 % v/v HNO$_3$; reference: H$_2$O)

7.2 Individual Overview Spectra

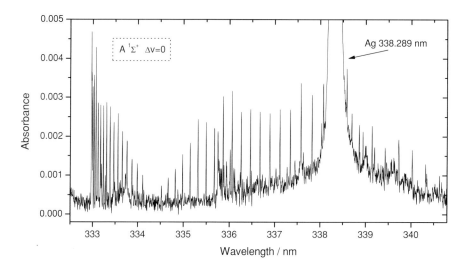

Figure 7.7: Spectrum detail of an air/acetylene flame showing the $\Delta v = 0$ sequence of the AgH band system around 336 nm; electronic transition $X\,^1\Sigma^+ \rightarrow A\,^1\Sigma^+$ (sample: 10 g/L Ag as AgNO$_3$ in 1 % v/v HNO$_3$; reference: H$_2$O)

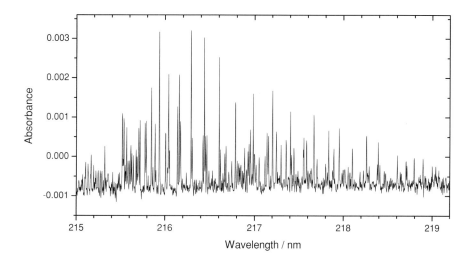

Figure 7.8: Spectrum detail of an air/acetylene flame with Ag, showing an *unknown* band system (possibly AgH) around 217 nm (sample: 10 g/L Ag as AgNO$_3$ in 1 % v/v HNO$_3$; reference: H$_2$O)

7.2.2 AlCl

In a nitrous oxide/acetylene flame Al produces structures of two diatomic molecules (refer to the overview spectrum in Figure 7.10). The longer wavelength structure is attributed to AlCl and results from a $\Delta v = 0$ sequence of the $X\,^1\Sigma^+ \rightarrow A\,^1\Pi$ electronic transition. Spectrum details are shown in Figure 7.9. Since the measurement of Al requires application of nitrous oxide, flame emissions become important and therefore the measurement was accomplished by making additionally use of the tungsten halogen lamp as well as the modified registration scheme to correct for flame emission. In total, 4×25 spectra were recorded with an illumination time of 10 s each. 5 g/L Al as $AlCl_3$ dissolved in 1 % v/v HNO_3 was used as the sample and deionized water as the reference. A list of characteristics of the corresponding AlCl bands is compiled in Table 7.2.

A further interesting detail can be seen in the overview spectrum of Figure 7.10 at 207 nm, which has already been discussed in Section 6.1. Absorbance measurements enable the investigation of highly excited atomic levels of aluminum (the so-called Rydberg states). In contrast to a registration in emission, where these states are not observable because of their quick depopulation due to the auto-ionization effects, the Rydberg series becomes clearly visible in absorbance (please refer also to Figure 6.1). The strength of the individual transitions and their spectral separation decrease towards the ionization limit and, looking carefully, two overlapping series are actually present.

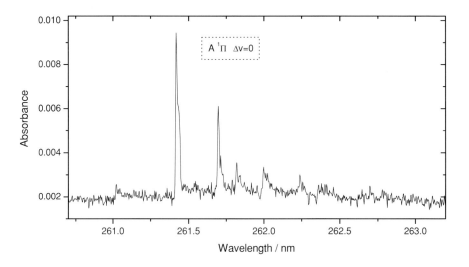

Figure 7.9: Spectrum detail of a nitous oxide/acetylene flame showing the $\Delta v = 0$ sequence of the AlCl band system around 262 nm; electronic transition $X\,^1\Sigma^+ \rightarrow A\,^1\Pi$ (sample: 5 g/L Al as $AlCl_3$ dissolved in 1 % v/v HNO_3; reference: deionized H_2O)

7.2 Individual Overview Spectra

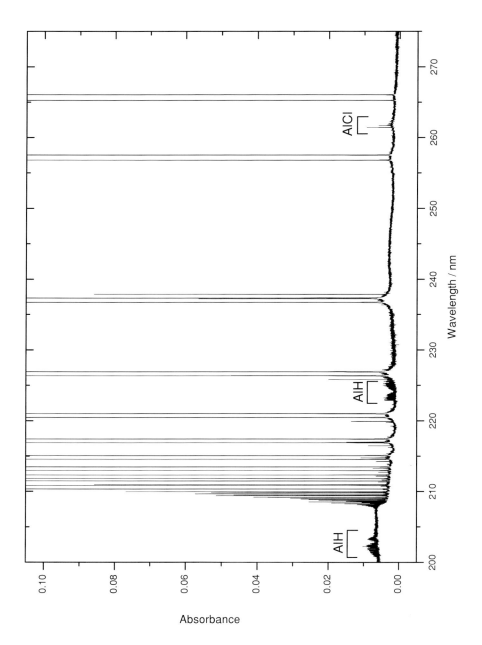

Figure 7.10: Overview spectrum of Al in a nitrous oxide / acetylene flame with bands of AlCl and AlH (sample: 5 g/L Al as $AlCl_3$ dissolved in 1 % v/v HNO_3; reference: deionized H_2O)

7. Electron Excitation Spectra of Diatomic Molecules

Table 7.2: Characteristics of the AlCl bands

Wavelength / nm	Relative intensity	Upper electronic energy / eV	Electronic transition	Vibrational transition
261.02	600	4.74	$X\,^1\Sigma^+ \rightarrow A\,^1\Pi$	$0 \rightarrow 0$
261.44	800	4.74	$X\,^1\Sigma^+ \rightarrow A\,^1\Pi$	$0 \rightarrow 0$
261.70	400	4.74	$X\,^1\Sigma^+ \rightarrow A\,^1\Pi$	$1 \rightarrow 1$
261.82	300	4.74	$X\,^1\Sigma^+ \rightarrow A\,^1\Pi$	$2 \rightarrow 2$
262.00	400	4.74	$X\,^1\Sigma^+ \rightarrow A\,^1\Pi$	$2 \rightarrow 2$
262.24	400	4.74	$X\,^1\Sigma^+ \rightarrow A\,^1\Pi$	$3 \rightarrow 3$
262.35	500	4.74	$X\,^1\Sigma^+ \rightarrow A\,^1\Pi$	$3 \rightarrow 3$
262.70	300	4.74	$X\,^1\Sigma^+ \rightarrow A\,^1\Pi$	$4 \rightarrow 4$

7.2.3 AlF

Using the standard air / acetylene flame and hydrofluoric acid a different diatomic aluminum molecule is formed. Now, AlF becomes visible, as seen in the overview spectrum of Figure 7.11 and the enlarged clipping in Figure 7.12. In comparison to the other two Al molecules measured in the nitrous oxide / acetylene flame, this band system is not so fine structured. It results from a $\Delta v = 0$ sequence of the $X\,^1\Sigma^+ \rightarrow A\,^1\Pi$ electronic transition around 227.5 nm.

For signal registration the standard measurement scheme was used, consisting of 2×40 recordings with an individual illumination time of 17 s. 0.1 g/L Al as $AlCl_3$ dissolved in 1 % v/v HNO_3 + 2 % v/v HF was used as the sample and deionized water as the reference. A list of characteristics of the AlF bands is compiled in Table 7.3.

Table 7.3: Characteristics of the AlF bands

Wavelength / nm	Relative intensity	Upper electronic energy / eV	Electronic transition	Vibrational transition
227.47	1000	5.45	$X\,^1\Sigma^+ \rightarrow A\,^1\Pi$	$0, 1 \rightarrow 0, 1$
227.49	800	5.45	$X\,^1\Sigma^+ \rightarrow A\,^1\Pi$	$2 \rightarrow 2$

7.2 Individual Overview Spectra

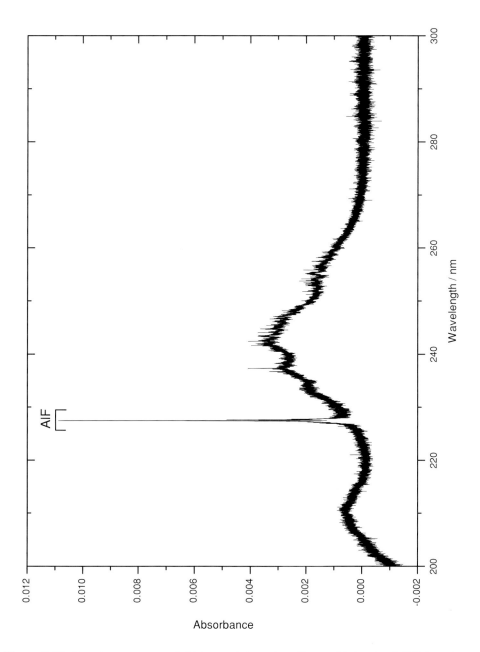

Figure 7.11: Overview spectrum of Al in an air/acetylene flame with bands of AlF (sample: 0.1 g/L Al as $AlCl_3$ dissolved in 1 % v/v HNO_3 + 2 % v/v HF; reference: deionized H_2O)

7. Electron Excitation Spectra of Diatomic Molecules

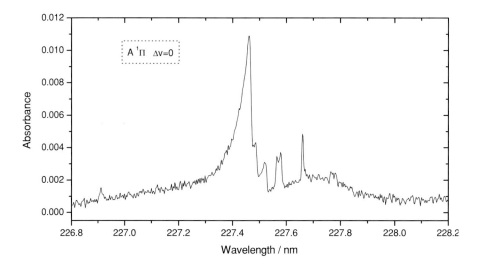

Figure 7.12: Spectrum detail of an air/acetylene flame showing the $\Delta v = 0$ sequence of the AlF band system around 227.5 nm; electronic transition $X\,^1\Sigma^+ \rightarrow A\,^1\Pi$ (sample: 0.1 g/L Al as $AlCl_3$ dissolved in 1 % v/v HNO_3 + 2 % v/v HF; reference: deionized H_2O)

7.2.4 AlH

As already indicated in Section 7.2.2 there is a third Al compound: AlH. Using the nitrous oxide/acetylene flame two AlH bands become visible (see Figure 7.10). The first band, shown in Figure 7.13, is due to the electronic transition to the $D\,^1\Sigma^+$ state around 202 nm, while for the second one, around 224 nm (Figure 7.14), the transition ends up in the $C\,^1\Sigma^+$ electronic state. Both transitions start from the same electronic ground state ($X\,^1\Sigma^+$) and

Table 7.4: Characteristics of the AlH bands

Wavelength / nm	Relative intensity	Upper electronic energy / eV	Electronic transition	Vibrational transition
202.82	?	6.11	$X\,^1\Sigma^+ \rightarrow D\,^1\Sigma^+$	$0 \rightarrow 0$
222.86	?	5.55	$X\,^1\Sigma^+ \rightarrow C\,^1\Sigma^+$	$0 \rightarrow 0$
224.16	?	5.55	$X\,^1\Sigma^+ \rightarrow C\,^1\Sigma^+$	$0 \rightarrow 0$
225.58	?	5.55	$X\,^1\Sigma^+ \rightarrow C\,^1\Sigma^+$	$1 \rightarrow 1$

7.2 *Individual Overview Spectra*

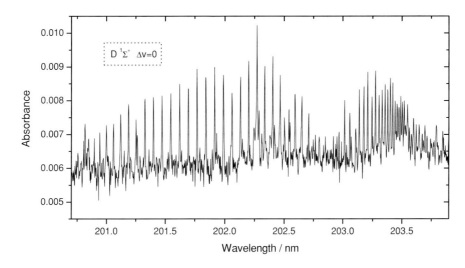

Figure 7.13: Spectrum detail of a nitrous oxide / acetylene flame showing the $\Delta v = 0$ sequence of the AlH band system around 202 nm; electronic transition $X\,^1\Sigma^+ \to D\,^1\Sigma^+$ (sample: 5 g/L Al as $AlCl_3$ dissolved in 1 % v/v HNO_3; reference: deionized H_2O)

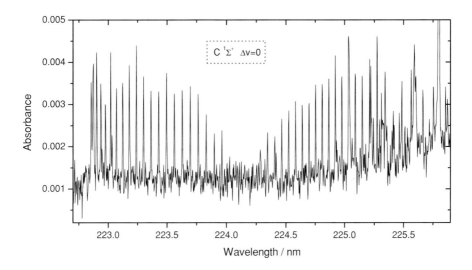

Figure 7.14: Spectrum detail of a nitrous oxide / acetylene flame showing the $\Delta v = 0$ sequence of the AlH band system around 224 nm; electronic transition $X\,^1\Sigma^+ \to C\,^1\Sigma^+$ (sample: 5 g/L Al as $AlCl_3$ dissolved in 1 % v/v HNO_3; reference: deionized H_2O)

7. Electron Excitation Spectra of Diatomic Molecules

comprise a $\Delta v = 0$ sequence. A compilation of the AlF band characteristics is listed in Table 7.4.

The signal registration scheme was similar to that used in Section 7.2.2 and even the solutions were the same, which means that both, AlH and AlCl, are simultaneously formed in the nitrous oxide / acetylene flame.

7.2.5 AsO

Arsenic, as the element with the shortest primary absorption wavelength in AAS, produces several AsO molecular bands around 250 nm if the conventional air / acetylene flame is used. An overview of the structures produced by the diatomic molecule is shown in Figure 7.15. Here the electronic transition takes place between the $X\,^2\Pi$ and the $B\,^2\Sigma^+$ state, while the vibrational excitation is changed in some cases. Apart from the two AsO bands with $\Delta v = +1$ around 243.5 nm and $\Delta v = -1$ around 256.3 nm, two systems without a change of the vibrational excitation exist. The latter can be divided into the so-called subsystems I and II, showing pronounced band heads around 250.4 nm and 257.0 nm, respectively. In Figure 7.16 the enlarged spectrum of the electronic transition to the $B\,^2\Sigma^+$ state with $\Delta v = 0$ (subsystems I) is depicted.

For signal registration the modified measurement scheme using the additional tungsten halogen lamp as well as the emission correction procedure was applied. 4×25 measurements with an illumination time of 12 s each were recorded. A 1 g/L As solution prepared from H_3AsO_4 dissolved in 3.5 % v/v HNO_3 was used as the sample and deionized water as the reference. The characteristics of AsO bands are listed in Table 7.5.

Table 7.5: Characteristics of the AsO bands

Wavelength / nm	Relative intensity	Upper electronic energy / eV	Electronic transition	Vibrational transition
243.85	300	4.94	$X\,^2\Pi \rightarrow B\,^2\Sigma^+$	$0 \rightarrow 1$
250.36	700	4.94	$X\,^2\Pi \rightarrow B\,^2\Sigma^+$	$0 \rightarrow 0$
250.47	600	4.94	$X\,^2\Pi \rightarrow B\,^2\Sigma^+$	$0 \rightarrow 0$
256.52	400	4.94	$X\,^2\Pi \rightarrow B\,^2\Sigma^+$	$1 \rightarrow 0$
256.97	800	4.94	$X\,^2\Pi \rightarrow B\,^2\Sigma^+$	$0 \rightarrow 0$
257.09	600	4.94	$X\,^2\Pi \rightarrow B\,^2\Sigma^+$	$0 \rightarrow 0$

7.2 Individual Overview Spectra

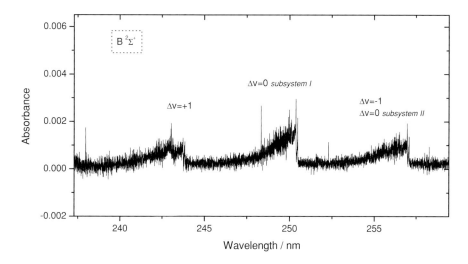

Figure 7.15: Enlarged overview spectrum of the AsO band system in an air/acetylene flame around 250 nm; electronic transition $X\,^2\Pi \rightarrow B\,^2\Sigma^+$ (sample: 1 g/L As as H_3AsO_4 dissolved in 3.5 % v/v HNO_3; reference: deionized H_2O)

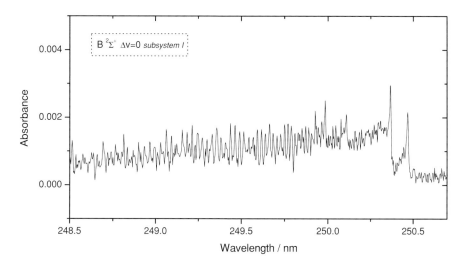

Figure 7.16: Spectrum detail of an air/acetylene flame showing the $\Delta v = 0$ (I) sequence of the AsO band system around 249.5 nm; electronic transition $X\,^2\Pi \rightarrow B\,^2\Sigma^+$ (sample: 1 g/L As as H_3AsO_4 dissolved in 3.5 % v/v HNO_3; reference: deionized H_2O)

7.2.6 CN

Due to the presence of carbon, hydrogen, nitrogen and oxygen, the nitrous oxide (N_2O)/ acetylene (C_2H_2) flame itself produces a lot of different diatomic molecule structures as can be seen in the overview spectrum of Figure 7.17. They are present all the time using this kind of flame and therefore special attention should be paid to them since they can increase the BG noise level.

One of these diatomic candidates is CN, of which three band systems are present, all lying in the longer wavelength region. They result from the same electronic transitions between the $X\,^2\Sigma^+$ and the $B\,^2\Sigma^+$ state. The strongest one, in the region of about 385 nm (see Figure 7.19), is related to a $\Delta v = 0$ sequence. The second system around 357 nm, as depicted in Figure 7.18, belongs to the transition increasing the vibrational excitation by a value of 1 ($\Delta v = +1$), while for the third and weakest one around 418 nm (refer to Figure 7.20) the excitation in vibration is reduced ($\Delta v = -1$). A list of the characteristics of the corresponding CN bands is compiled in Table 7.6.

The nitrous oxide/acetylene flame is well known for its strong emissions and thus the modified registration scheme was applied to record the spectra. In total, 4×50 spectra have been recorded with an individual illumination time of 12 s. For the sample measurements the nitrous oxide/acetylene flame was turned on, while in the case of blank signal registration the flame was turned off. No additional solution was aspirated by the nebulizer to get the pure spectrum of the flame gases.

Table 7.6: Characteristics of the CN bands

Wavelength / nm	Relative intensity	Upper electronic energy / eV	Electronic transition	Vibrational transition
358.39	600	3.19	$X\,^2\Sigma^+ \to B\,^2\Sigma^+$	$2 \to 3$
358.59	700	3.19	$X\,^2\Sigma^+ \to B\,^2\Sigma^+$	$1 \to 2$
359.04	800	3.19	$X\,^2\Sigma^+ \to B\,^2\Sigma^+$	$0 \to 1$
386.19	700	3.19	$X\,^2\Sigma^+ \to B\,^2\Sigma^+$	$2 \to 2$
387.14	900	3.19	$X\,^2\Sigma^+ \to B\,^2\Sigma^+$	$1 \to 1$
388.34	1000	3.19	$X\,^2\Sigma^+ \to B\,^2\Sigma^+$	$0 \to 0$
418.10	700	3.19	$X\,^2\Sigma^+ \to B\,^2\Sigma^+$	$3 \to 2$
419.72	800	3.19	$X\,^2\Sigma^+ \to B\,^2\Sigma^+$	$2 \to 1$
421.60	900	3.19	$X\,^2\Sigma^+ \to B\,^2\Sigma^+$	$1 \to 0$

7.2 Individual Overview Spectra

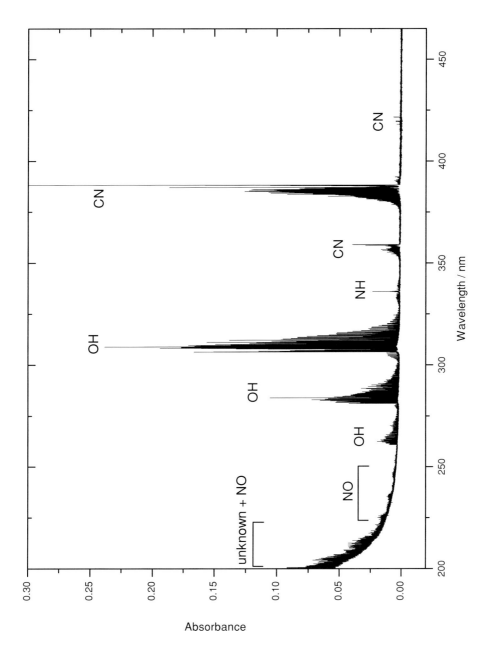

Figure 7.17: Overview spectrum of a nitrous oxide / acetylene flame with bands of NO, OH, NH, CN, and an unknown molecule (sample: flame turned on; reference: flame turned off)

7. Electron Excitation Spectra of Diatomic Molecules

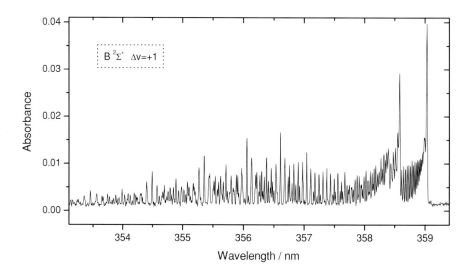

Figure 7.18: Spectrum detail of a nitrous oxide / acetylene flame showing the $\Delta v = +1$ sequence of the CN band system around 357 nm; electronic transition $X\,^2\Sigma^+ \rightarrow B\,^2\Sigma^+$ (sample: flame turned on; reference: flame turned off)

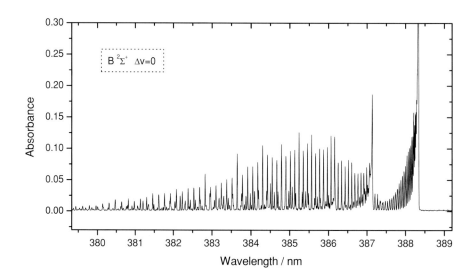

Figure 7.19: Spectrum detail of a nitrous oxide / acetylene flame showing the $\Delta v = 0$ sequence of the CN band system around 385 nm; electronic transition $X\,^2\Sigma^+ \rightarrow B\,^2\Sigma^+$ (sample: flame turned on; reference: flame turned off)

7.2 Individual Overview Spectra

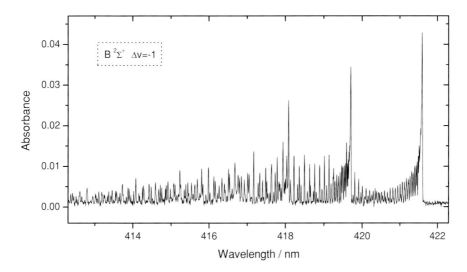

Figure 7.20: Spectrum detail of a nitrous oxide/acetylene flame showing the $\Delta v = -1$ sequence of the CN band system around 418 nm; electronic transition $X\,^2\Sigma^+ \rightarrow B\,^2\Sigma^+$ (sample: flame turned on; reference: flame turned off)

The second diatomic molecule visible in the nitrous oxide/acetylene flame is NH which will be found in the wavelength region around 336 nm. The NH structure is discussed later in Section 7.2.11. Moreover, NO structures are also produced by the nitrous oxide/acetylene flame. They start at a wavelength of about 250 nm and go down to the short wavelength limit of the ARES spectrograph. Here they are overlapped by another not yet identified structure. Of course, NO bands are more efficiently generated with the use of nitric acid as sample solution and will be discussed later in Section 7.2.12.

Last but not least, strong OH structures are also present. They are not limited to this flame type since the conventional air/acetylene flame also generates these molecule bands. For further details refer to Section 7.2.13.

7.2.7 CS

Using sulfuric acid (H_2SO_4) as the sample solution, CS bands are produced in an air/acetylene flame, as shown in Figure 7.21. The main band at about 258.5 nm, as depicted in Figure 7.22, results from the electronic transition between the $X\,^1\Sigma^+$ and the $A\,^1\Pi$ state, again with a $\Delta v = 0$ sequence. In Figure 7.23, the sequences with increasing vibrational excitation ($\Delta v = +1, 2, \ldots$) can be found to the left, while the decreasing ones are shifted to longer wavelengths.

7. Electron Excitation Spectra of Diatomic Molecules

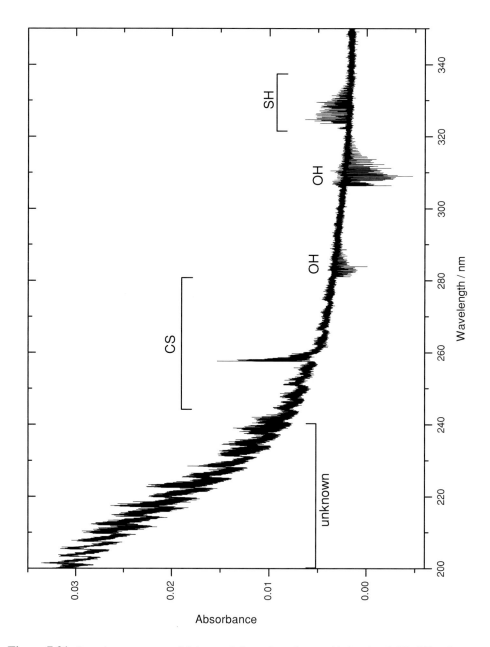

Figure 7.21: Overview spectrum of S in an air/acetylene flame with bands of CS, SH and an unknown molecule (sample: 5 % v/v H_2SO_4; reference: deionized H_2O)

7.2 Individual Overview Spectra

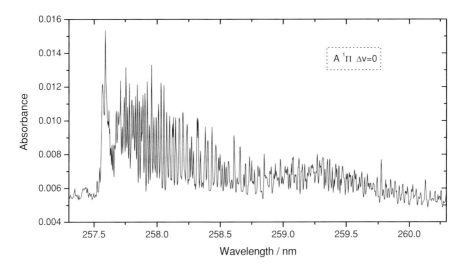

Figure 7.22: Spectrum detail of an air/acetylene flame showing the $\Delta v = 0$ sequence of the CS band system around 258.5 nm; electronic transition $X\,^1\Sigma^+ \rightarrow A\,^1\Pi$ (sample: 5 % v/v H_2SO_4; reference: deionized H_2O)

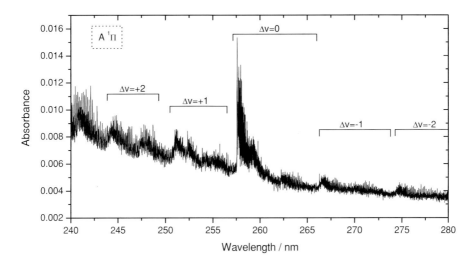

Figure 7.23: Enlarged overview spectrum of the CS band system in an air/acetylene flame around 260 nm; electronic transition $X\,^1\Sigma^+ \rightarrow A\,^1\Pi$ (sample: 5 % v/v H_2SO_4; reference: deionized H_2O)

7. Electron Excitation Spectra of Diatomic Molecules

The standard measurement scheme was used for signal registration, consisting of 2×50 individual recordings with an illumination time of 17 s each. A 5 % v/v H_2SO_4 solution was used as the sample and deionized water as the reference. A compilation of the characteristics of the CS bands can be found in Table 7.7.

Table 7.7: Characteristics of the CS bands

Wavelength / nm	Relative intensity	Upper electronic energy / eV	Electronic transition	Vibrational transition
244.48	300	4.83	$X\,^1\Sigma^+ \to A\,^1\Pi$	$0 \to 2$
246.02	500	4.83	$X\,^1\Sigma^+ \to A\,^1\Pi$	$1 \to 3$
247.70	400	4.83	$X\,^1\Sigma^+ \to A\,^1\Pi$	$2 \to 4$
250.73	400	4.83	$X\,^1\Sigma^+ \to A\,^1\Pi$	$0 \to 1$
252.32	700	4.83	$X\,^1\Sigma^+ \to A\,^1\Pi$	$1 \to 2$
253.87	800	4.83	$X\,^1\Sigma^+ \to A\,^1\Pi$	$2 \to 3$
255.58	500	4.83	$X\,^1\Sigma^+ \to A\,^1\Pi$	$3 \to 4$
257.27	500	4.83	$X\,^1\Sigma^+ \to A\,^1\Pi$	$4 \to 5$
257.56	1000	4.83	$X\,^1\Sigma^+ \to A\,^1\Pi$	$0 \to 0$
258.96	600	4.83	$X\,^1\Sigma^+ \to A\,^1\Pi$	$1 \to 1$
260.59	800	4.83	$X\,^1\Sigma^+ \to A\,^1\Pi$	$2 \to 2$
262.16	700	4.83	$X\,^1\Sigma^+ \to A\,^1\Pi$	$3 \to 3$
263.89	200	4.83	$X\,^1\Sigma^+ \to A\,^1\Pi$	$4 \to 4$
266.26	900	4.83	$X\,^1\Sigma^+ \to A\,^1\Pi$	$1 \to 0$
267.70	600	4.83	$X\,^1\Sigma^+ \to A\,^1\Pi$	$2 \to 1$
269.32	700	4.83	$X\,^1\Sigma^+ \to A\,^1\Pi$	$3 \to 2$
270.89	600	4.83	$X\,^1\Sigma^+ \to A\,^1\Pi$	$4 \to 3$
272.67	400	4.83	$X\,^1\Sigma^+ \to A\,^1\Pi$	$5 \to 4$
274.39	300	4.83	$X\,^1\Sigma^+ \to A\,^1\Pi$	$6 \to 5$
275.47	700	4.83	$X\,^1\Sigma^+ \to A\,^1\Pi$	$2 \to 0$
276.92	300	4.83	$X\,^1\Sigma^+ \to A\,^1\Pi$	$3 \to 1$

In addition to the CS structures, SH bands are also observed in the air/acetylene flame aspirating sulfuric acid. They can be observed in the wavelength region around 326 nm. Further details about the SH molecule structures are given in Section 7.2.15. For wavelengths below 240 nm other very strong molecular structures appear that cannot be classified yet.

7.2.8 CuH

When investigating copper, a hydride is formed in the conventional air/acetylene flame. An overview is given in Figure 7.24 showing some Cu atomic lines, too. The CuH bands in the shorter wavelength region around 224 nm result from a $\Delta v = 0$ sequence of the $X\,^1\Sigma^+ \to D\,^1\Pi$ electronic transition. They are partly overlapped by much weaker NO structures caused by the HNO_3 solvent, lying in the same wavelength region, so that the spectrum shown in Figure 7.25 consists, to a certain extent, of a mixture of both. The other strong CuH bands appear at wavelengths of about 435 nm (refer to Figure 7.26). Here the corresponding electronic transition takes place between the $X\,^1\Sigma^+$ and the $A\,^1\Sigma^+$ state, again with ($\Delta v = 0$). The transitions to the $B\,^1\Sigma^+$ state, as mentioned in the literature, are not visible. The characteristics of the CuH bands are listed in Table 7.8.

Table 7.8: Characteristics of the CuH bands

Wavelength / nm	Relative intensity	Upper electronic energy / eV	Electronic transition	Vibrational transition
222.80	?	5.54	$X\,^1\Sigma^+ \to D\,^1\Pi$	$0 \to 0$
223.40	?	5.54	$X\,^1\Sigma^+ \to D\,^1\Pi$	$1 \to 1$
223.92	?	5.54	$X\,^1\Sigma^+ \to D\,^1\Pi$	$0 \to 0$
224.20	?	5.54	$X\,^1\Sigma^+ \to D\,^1\Pi$	$1 \to 1$
400.54	?	2.91	$X\,^1\Sigma^+ \to A\,^1\Sigma^+$	$0 \to 1$
401.15	?	2.91	$X\,^1\Sigma^+ \to A\,^1\Sigma^+$	$0 \to 1$
406.19	?	2.91	$X\,^1\Sigma^+ \to A\,^1\Sigma^+$	$1 \to 2$
406.76	?	2.91	$X\,^1\Sigma^+ \to A\,^1\Sigma^+$	$1 \to 2$
408.74	?	3.28	$X\,^1\Sigma^+ \to B\,^1\Sigma^+$	$1 \to 0$
409.46	?	3.28	$X\,^1\Sigma^+ \to B\,^1\Sigma^+$	$1 \to 0$
412.74	?	3.28	$X\,^1\Sigma^+ \to B\,^1\Sigma^+$	$2 \to 1$
413.35	?	3.28	$X\,^1\Sigma^+ \to B\,^1\Sigma^+$	$2 \to 1$
427.96	?	2.91	$X\,^1\Sigma^+ \to A\,^1\Sigma^+$	$0 \to 0$
428.86	?	2.91	$X\,^1\Sigma^+ \to A\,^1\Sigma^+$	$0 \to 0$
432.77	?	2.91	$X\,^1\Sigma^+ \to A\,^1\Sigma^+$	$1 \to 1$
433.62	?	2.91	$X\,^1\Sigma^+ \to A\,^1\Sigma^+$	$1 \to 1$
437.90	?	2.91	$X\,^1\Sigma^+ \to A\,^1\Sigma^+$	$2 \to 2$
438.74	?	2.91	$X\,^1\Sigma^+ \to A\,^1\Sigma^+$	$2 \to 2$
464.84	?	2.91	$X\,^1\Sigma^+ \to A\,^1\Sigma^+$	$1 \to 0$

7. Electron Excitation Spectra of Diatomic Molecules

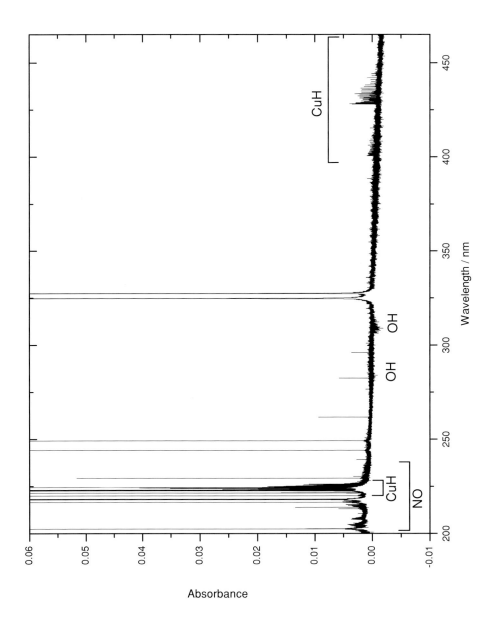

Figure 7.24: Overview spectrum of Cu in an air / acetylene flame with bands of CuH and NO (sample: 10 g/L Cu as $Cu(NO_3)_2$ in 2 % v/v HNO_3; reference: deionized H_2O)

7.2 Individual Overview Spectra

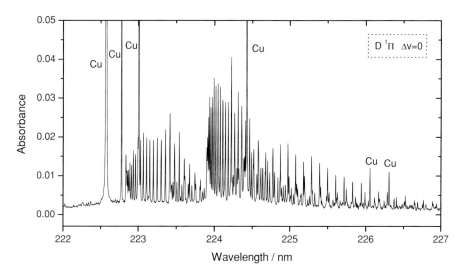

Figure 7.25: Spectrum detail of an air/acetylene flame showing the $\Delta v = 0$ sequence of the CuH band system around 224 nm; electronic transition $X\ ^1\Sigma^+ \rightarrow D\ ^1\Pi$ (sample: 10 g/L Cu as Cu(NO$_3$)$_2$ in 2 % v/v HNO$_3$; reference: deionized H$_2$O)

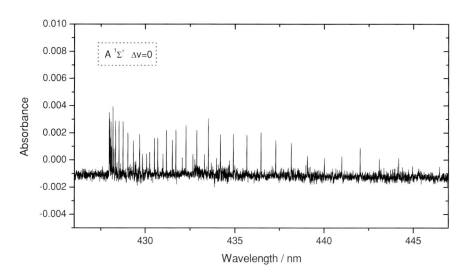

Figure 7.26: Spectrum detail of an air/acetylene flame showing the $\Delta v = 0$ sequence of the CuH band system around 435 nm; electronic transition $X\ ^1\Sigma^+ \rightarrow A\ ^1\Sigma^+$ (sample: 10 g/L Cu as Cu(NO$_3$)$_2$ in 2 % v/v HNO$_3$; reference: deionized H$_2$O)

7. Electron Excitation Spectra of Diatomic Molecules

For measurement of the CuH absorbance above 400 nm, due to the longer wavelengths to be investigated, the tungsten halogen lamp and the emission correction scheme were additionally used for signal recording. 4×50 measurements with an illumination time of 12 s each were recorded. A 10 g/L Cu solution, prepared from $Cu(NO_3)_2$ dissolved in 2 % v/v HNO_3, was used as the sample and deionized water as the reference.

7.2.9 GaCl

The shape of the structures produced by the GaCl molecule, as depicted in Figure 7.27, are eye-catchingly different to those of other diatomic molecules since they are not that fine structured. Nevertheless, the positions of the three bands (see Table 7.9), all resulting from the electronic transition between the $X\,^1\Sigma^+$ and the $C\,^1\Pi(1)$ state, can be fully related to the spectrum. Unfortunately, apart from the Ga atomic lines and an unknown molecule band system, the GaCl structures are overlapped by a huge broad-band structure, better seen in the overview spectrum of Figure 7.28. Its origin is still unknown but because of the smoothness it must result from a polyatomic molecule present in the flame.

Signal registration was accomplished using the standard measurement scheme, consisting of 2×50 recordings with an illumination time of 17 s each. 10 g/L Ga as $Ga(NO_3)_3$ dissolved in 2 % v/v HNO_3 / HCl was used as sample and deionized water as reference.

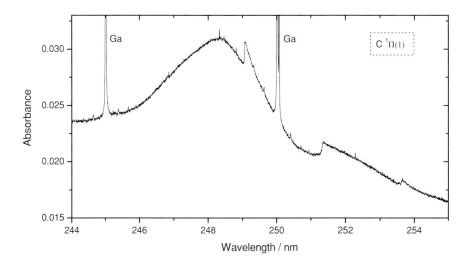

Figure 7.27: Spectrum detail of an air / acetylene flame showing some sequences of the GaCl band system around 249 nm; electronic transition $X\,^1\Sigma^+ \rightarrow C\,^1\Pi(1)$ (sample: 10 g/L Ga as $Ga(NO_3)_3$ in 2 % v/v HNO_3 / HCl; reference: deionized H_2O)

7.2 Individual Overview Spectra

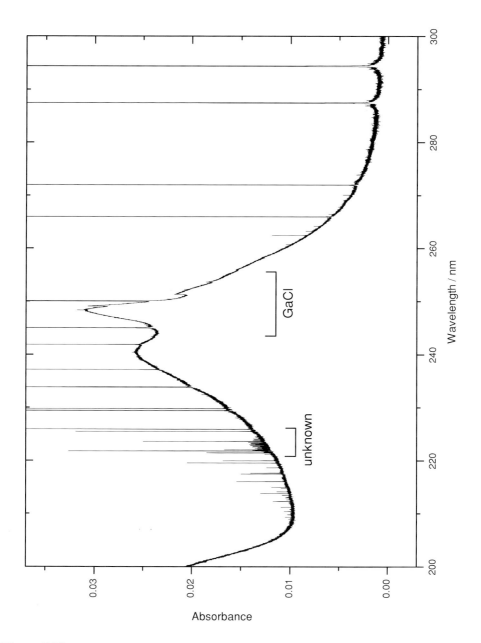

Figure 7.28: Overview spectrum of Ga in an air/acetylene flame with bands of GaCl and an unknown molecule (sample: 10 g/L Ga as Ga(NO$_3$)$_3$ in 2 % v/v HNO$_3$/HCl; reference: deionized H$_2$O)

7. Electron Excitation Spectra of Diatomic Molecules

Table 7.9: Characteristics of the GaCl bands

Wavelength / nm	Relative intensity	Upper electronic energy / eV	Electronic transition	Vibrational transition
249.06	1000	4.99	$X\,^1\Sigma^+ \to C\,^1\Pi(1)$	$0 \to 0$
251.33	800	4.99	$X\,^1\Sigma^+ \to C\,^1\Pi(1)$	$1 \to 0$
253.65	600	4.99	$X\,^1\Sigma^+ \to C\,^1\Pi(1)$	$2 \to 0$

7.2.10 LaO

The rare-earth element lanthanum produces molecule structures in the long wavelength region around 442 nm. Unfortunately, no spectral data are available in the literature regarding diatomic molecule bands generated by La. Thus, most likely the structures in Figure 7.29 are due to the formation of LaO, since a solution of 5 g/L La as La_2O_3 dissolved in 2 % v/v HCl was used for the registration.

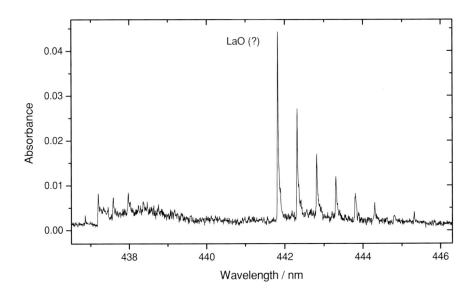

Figure 7.29: Enlarged overview spectrum of a likely LaO band system in an air/acetylene flame around 442 nm; electronic transition unknown (sample: 5 g/L La as La_2O_3 dissolved in 2 % v/v HCl; reference: deionized H_2O)

Because of the use of the nitrous oxide/acetylene flame the modified measurement scheme with additional tungsten halogen lamp and emission correction was applied. The registration cycle consisted of 4×25 recordings with an individual illumination time of 12 s.

7.2.11 NH

It has already been mentioned in Section 7.2.6 that the nitrous oxide/acetylene flame itself, apart from CN and NO bands, produces NH molecule structures. They are present in the wavelength region around 336 nm (refer to overview spectrum in Figure 7.17 and the enlarged clipping in Figure 7.30). The corresponding electronic transition with a $\Delta v = 0$ sequence connects the $X\,^3\Sigma^-$ to the $A\,^3\Pi$ state.

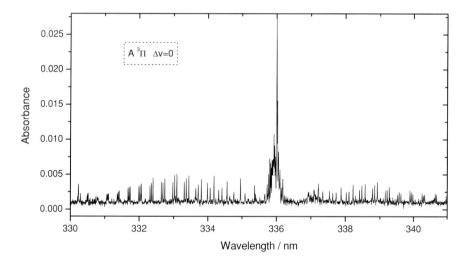

Figure 7.30: Spectrum detail of a nitrous oxide/acetylene flame showing the $\Delta v = 0$ sequence of the NH band system around 336 nm; electronic transition $X\,^3\Sigma^- \rightarrow A\,^3\Pi$ (sample: flame turned on; reference: flame turned off)

A compilation of the characteristics of the NH bands can be found in Table 7.10. For recording the spectra the modified registration scheme was applied to eliminate the strong emissions coming from the nitrous oxide/acetylene flame. In total, 4×50 spectra were recorded with an illumination time of 12 s each. In the case of sample measurements the nitrous oxide/acetylene flame was turned on, while in the case of blank signal registration the flame was turned off. No additional solution was aspirated by the nebulizer to get the pure spectrum of the flame gases.

Table 7.10: Characteristics of the NH bands

Wavelength / nm	Relative intensity	Upper electronic energy / eV	Electronic transition	Vibrational transition
330.25	400	3.69	$X\,^3\Sigma^- \to A\,^3\Pi$	$0 \to 0$
331.70	300	3.69	$X\,^3\Sigma^- \to A\,^3\Pi$	$1 \to 1$
336.01	1000	3.69	$X\,^3\Sigma^- \to A\,^3\Pi$	$0 \to 0$
337.00	800	3.69	$X\,^3\Sigma^- \to A\,^3\Pi$	$1 \to 1$
338.30	300	3.69	$X\,^3\Sigma^- \to A\,^3\Pi$	$2 \to 2$

7.2.12 NO

NO structures are best generated using nitric acid (HNO_3) and a standard air/acetylene flame. In this way, the NO bands are obtained in their pure form without any overlap of other unknown structures as in the case of the nitrous oxide/acetylene flame (refer to Section 7.2.6). From the overview spectrum of Figure 7.31 five groups of bands become visible, all resulting from the same electronic transition between the $X\,^2\Pi$ and the $A\,^2\Sigma^+$ state but with different changes in the vibrational excitation. Enlarged clippings of the four strongest NO structures are depicted in Figures 7.32–7.36. The one with the shortest wavelength in the region of about 204 nm increases the vibrational excitation by a value of 2. Going to longer wavelengths one will find in sequence the transitions with a change of +1 (around 214 nm), 0 (around 225 nm) and −1 (around 235 nm) in vibrational excitation, respectively.

A list of the characteristics of all the NO bands is compiled in Table 7.11. In this table another electronic transition which connects the $X\,^2\Pi$ ground state to the $B\,^2\Pi$ state is mentioned. The corresponding bands are, of course, too weak to been seen in the spectra. This is quite similar to the situation already mentioned for the CuH molecule (refer to Section 7.2.8).

For signal registration the standard measurement scheme was used, which consisted of 2×50 recordings with an individual illumination time of 10 s. A 5 % v/v HNO_3 solution was used as the sample and deionized water as the reference.

In addition to the NO structures NH bands are also moderately generated in the conventional air/acetylene flame. They are already known from the previous section (please refer to Figure 7.30) where they have been recorded with a much better SNR.

7.2 *Individual Overview Spectra*

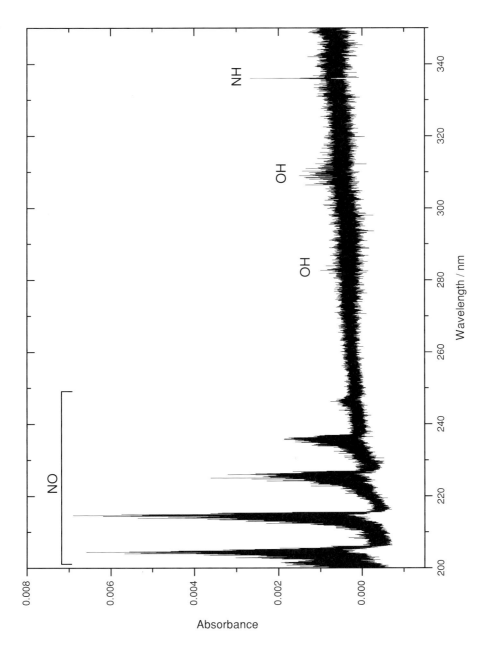

Figure 7.31: Overview spectrum of N in an air / acetylene flame with bands of NO and NH (sample: 5 % v/v HNO_3; reference: deionized H_2O)

7. Electron Excitation Spectra of Diatomic Molecules

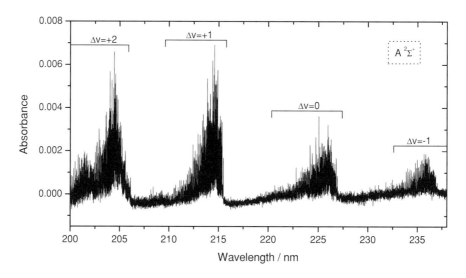

Figure 7.32: Enlarged overview spectrum of the NO band system in an air/acetylene flame around 220 nm; electronic transition $X\,^2\Pi \to A\,^2\Sigma^+$ (sample: 5 % v/v HNO$_3$; reference: deionized H$_2$O)

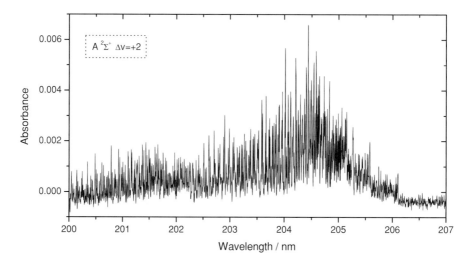

Figure 7.33: Spectrum detail of an air/acetylene flame showing the $\Delta v = +2$ sequence of the NO band system around 204 nm; electronic transition $X\,^2\Pi \to A\,^2\Sigma^+$ (sample: 5 % v/v HNO$_3$; reference: deionized H$_2$O)

7.2 Individual Overview Spectra

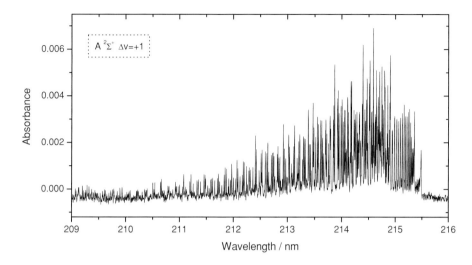

Figure 7.34: Spectrum detail of an air/acetylene flame showing the $\Delta v = +1$ sequence of the NO band system around 214 nm; electronic transition $X\,^2\Pi \rightarrow A\,^2\Sigma^+$ (sample: 5 % v/v HNO_3; reference: deionized H_2O)

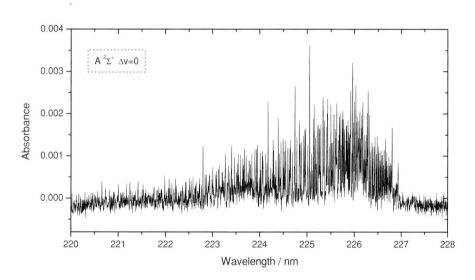

Figure 7.35: Spectrum detail of an air/acetylene flame showing the $\Delta v = 0$ sequence of the NO band system around 225 nm; electronic transition $X\,^2\Pi \rightarrow A\,^2\Sigma^+$ (sample: 5 % v/v HNO_3; reference: deionized H_2O)

183

7. Electron Excitation Spectra of Diatomic Molecules

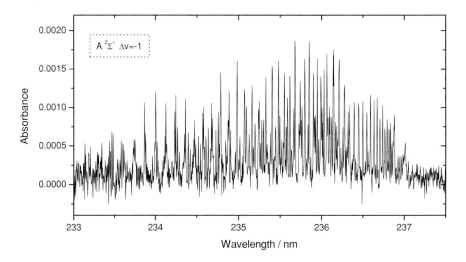

Figure 7.36: Spectrum detail of an air/acetylene flame showing the $\Delta v = -1$ sequence of the NO band system around 235 nm; electronic transition $X\,^2\Pi \rightarrow A\,^2\Sigma^+$ (sample: 5 % v/v HNO_3; reference: deionized H_2O)

Table 7.11: Characteristics of the NO bands

Wavelength / nm	Relative intensity	Upper electronic energy / eV	Electronic transition	Vibrational transition
203.07	200	5.45	$X\,^2\Pi \rightarrow A\,^2\Sigma^+$	$1 \rightarrow 3$
203.57	200	5.45	$X\,^2\Pi \rightarrow A\,^2\Sigma^+$	$1 \rightarrow 3$
204.75	400	5.45	$X\,^2\Pi \rightarrow A\,^2\Sigma^+$	$0 \rightarrow 2$
205.28	400	5.45	$X\,^2\Pi \rightarrow A\,^2\Sigma^+$	$0 \rightarrow 2$
212.96	100	5.45	$X\,^2\Pi \rightarrow A\,^2\Sigma^+$	$1 \rightarrow 2$
213.50	100	5.45	$X\,^2\Pi \rightarrow A\,^2\Sigma^+$	$1 \rightarrow 2$
214.91	700	5.45	$X\,^2\Pi \rightarrow A\,^2\Sigma^+$	$0 \rightarrow 1$
215.49	700	5.45	$X\,^2\Pi \rightarrow A\,^2\Sigma^+$	$0 \rightarrow 1$
219.40	100	5.45	$X\,^2\Pi \rightarrow A\,^2\Sigma^+$	$3 \rightarrow 3$
219.96	100	5.45	$X\,^2\Pi \rightarrow A\,^2\Sigma^+$	$3 \rightarrow 3$
221.63	300	5.45	$X\,^2\Pi \rightarrow A\,^2\Sigma^+$	$2 \rightarrow 2$

Table 7.11 (continued)

Wavelength / nm	Relative intensity	Upper electronic energy / eV	Electronic transition	Vibrational transition
222.24	300	5.45	$X\,^2\Pi \to A\,^2\Sigma^+$	$2 \to 2$
223.20	600	5.69	$X\,^2\Pi \to B\,^2\Pi$	$2 \to 3$
223.61	600	5.69	$X\,^2\Pi \to B\,^2\Pi$	$2 \to 3$
223.94	300	5.45	$X\,^2\Pi \to A\,^2\Sigma^+$	$1 \to 1$
224.54	300	5.45	$X\,^2\Pi \to A\,^2\Sigma^+$	$1 \to 1$
226.28	800	5.45	$X\,^2\Pi \to A\,^2\Sigma^+$	$0 \to 0$
226.94	800	5.45	$X\,^2\Pi \to A\,^2\Sigma^+$	$0 \to 0$
228.76	200	5.69	$X\,^2\Pi \to B\,^2\Pi$	$2 \to 2$
230.95	200	5.45	$X\,^2\Pi \to A\,^2\Sigma^+$	$3 \to 2$
231.63	200	5.45	$X\,^2\Pi \to A\,^2\Sigma^+$	$3 \to 2$
232.66	300	5.69	$X\,^2\Pi \to B\,^2\Pi$	$3 \to 3$
233.14	300	5.69	$X\,^2\Pi \to B\,^2\Pi$	$3 \to 3$
236.33	1000	5.45	$X\,^2\Pi \to A\,^2\Sigma^+$	$1 \to 0$
237.02	1000	5.45	$X\,^2\Pi \to A\,^2\Sigma^+$	$1 \to 0$
244.00	300	5.45	$X\,^2\Pi \to A\,^2\Sigma^+$	$3 \to 1$
244.70	300	5.45	$X\,^2\Pi \to A\,^2\Sigma^+$	$3 \to 1$
247.11	1000	5.45	$X\,^2\Pi \to A\,^2\Sigma^+$	$2 \to 0$
247.87	1000	5.45	$X\,^2\Pi \to A\,^2\Sigma^+$	$2 \to 0$

7.2.13 OH

OH is the diatomic molecule present everywhere in flame AAS, independent of the type of gases used. It generates huge fine-structured bands separated into four groups, as can be seen in the overview spectrum of Figure 7.37. They are due to the electronic transition between the $X\,^2\Pi$ and the $A\,^2\Sigma^+$ state. The strongest structure (refer to Figure 7.38) lying in the wavelength region around 310 nm does not change the vibrational excitation ($\Delta v = 0$). Towards shorter wavelengths, as shown in the enlarged clippings of Figures 7.39 and 7.40, one will find the transitions with increasing change in vibrational excitation ($\Delta v = +1, +2, \ldots$). The fourth band system at a wavelength of about 245 nm is not depicted in detail since the SNR was not favorable. A compilation of the characteristics of the OH bands is given in Table 7.12.

Table 7.12: Characteristics of the OH bands

Wavelength / nm	Relative intensity	Upper electronic energy / eV	Electronic transition	Vibrational transition
244.40	?	4.05	$X\,^2\Pi \rightarrow A\,^2\Sigma^+$	$0 \rightarrow 3$
260.85	300	4.05	$X\,^2\Pi \rightarrow A\,^2\Sigma^+$	$0 \rightarrow 2$
261.34	300	4.05	$X\,^2\Pi \rightarrow A\,^2\Sigma^+$	$0 \rightarrow 2$
262.21	300	4.05	$X\,^2\Pi \rightarrow A\,^2\Sigma^+$	$0 \rightarrow 2$
267.73	200	4.05	$X\,^2\Pi \rightarrow A\,^2\Sigma^+$	$1 \rightarrow 3$
268.18	200	4.05	$X\,^2\Pi \rightarrow A\,^2\Sigma^+$	$1 \rightarrow 3$
268.31	200	4.05	$X\,^2\Pi \rightarrow A\,^2\Sigma^+$	$1 \rightarrow 3$
269.11	200	4.05	$X\,^2\Pi \rightarrow A\,^2\Sigma^+$	$1 \rightarrow 3$
281.13	600	4.05	$X\,^2\Pi \rightarrow A\,^2\Sigma^+$	$0 \rightarrow 1$
281.60	600	4.05	$X\,^2\Pi \rightarrow A\,^2\Sigma^+$	$0 \rightarrow 1$
281.91	600	4.05	$X\,^2\Pi \rightarrow A\,^2\Sigma^+$	$0 \rightarrow 1$
282.90	600	4.05	$X\,^2\Pi \rightarrow A\,^2\Sigma^+$	$0 \rightarrow 1$
287.53	300	4.05	$X\,^2\Pi \rightarrow A\,^2\Sigma^+$	$1 \rightarrow 2$
288.06	300	4.05	$X\,^2\Pi \rightarrow A\,^2\Sigma^+$	$1 \rightarrow 2$
288.23	300	4.05	$X\,^2\Pi \rightarrow A\,^2\Sigma^+$	$1 \rightarrow 2$
289.27	300	4.05	$X\,^2\Pi \rightarrow A\,^2\Sigma^+$	$1 \rightarrow 2$
294.52	100	4.05	$X\,^2\Pi \rightarrow A\,^2\Sigma^+$	$2 \rightarrow 3$
295.12	100	4.05	$X\,^2\Pi \rightarrow A\,^2\Sigma^+$	$2 \rightarrow 3$
296.24	100	4.05	$X\,^2\Pi \rightarrow A\,^2\Sigma^+$	$2 \rightarrow 3$
302.12	1000	4.05	$X\,^2\Pi \rightarrow A\,^2\Sigma^+$	$0 \rightarrow 0$
306.36	1000	4.05	$X\,^2\Pi \rightarrow A\,^2\Sigma^+$	$0 \rightarrow 0$
306.72	1000	4.05	$X\,^2\Pi \rightarrow A\,^2\Sigma^+$	$0 \rightarrow 0$
307.80	1000	4.05	$X\,^2\Pi \rightarrow A\,^2\Sigma^+$	$0 \rightarrow 0$
308.90	1000	4.05	$X\,^2\Pi \rightarrow A\,^2\Sigma^+$	$0 \rightarrow 0$
312.17	100	4.05	$X\,^2\Pi \rightarrow A\,^2\Sigma^+$	$1 \rightarrow 1$
312.64	100	4.05	$X\,^2\Pi \rightarrow A\,^2\Sigma^+$	$1 \rightarrow 1$
318.48	100	4.05	$X\,^2\Pi \rightarrow A\,^2\Sigma^+$	$2 \rightarrow 2$
319.02	100	4.05	$X\,^2\Pi \rightarrow A\,^2\Sigma^+$	$2 \rightarrow 2$
319.59	100	4.05	$X\,^2\Pi \rightarrow A\,^2\Sigma^+$	$2 \rightarrow 2$
320.87	100	4.05	$X\,^2\Pi \rightarrow A\,^2\Sigma^+$	$2 \rightarrow 2$

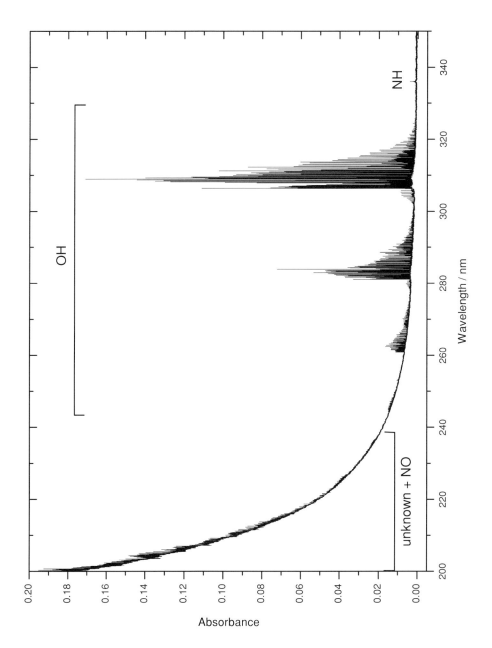

Figure 7.37: Overview spectrum of an air/acetylene flame with bands of OH, NH, NO and an unknown molecule (sample: flame turned on; reference: flame turned off)

7. *Electron Excitation Spectra of Diatomic Molecules*

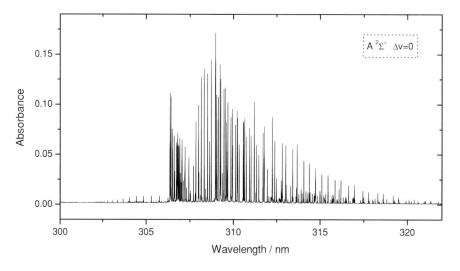

Figure 7.38: Spectrum detail of an air/acetylene flame showing the $\Delta v = 0$ sequence of the OH band system around 310 nm; electronic transition $X\,^2\Pi \rightarrow A\,^2\Sigma^+$ (sample: flame turned on; reference: flame turned off)

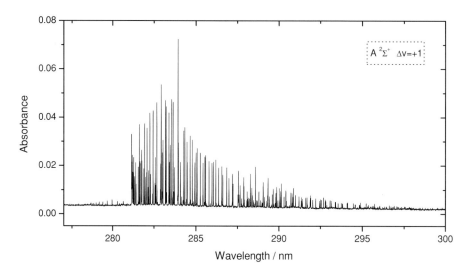

Figure 7.39: Spectrum detail of an air/acetylene flame showing the $\Delta v = +1$ sequence of the OH band system around 285 nm; electronic transition $X\,^2\Pi \rightarrow A\,^2\Sigma^+$ (sample: flame turned on; reference: flame turned off)

7.2 Individual Overview Spectra

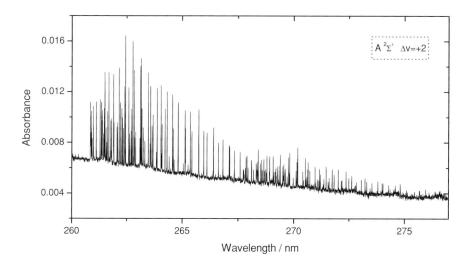

Figure 7.40: Spectrum detail of an air/acetylene flame showing the $\Delta v = +2$ sequence of the OH band system around 265 nm; electronic transition $X\,^2\Pi \rightarrow A\,^2\Sigma^+$ (sample: flame turned on; reference: flame turned off)

Fortunately, the OH bands are normally not a general problem for conventional BC systems since they are present during sample as well as blank measurement and are of the same order on average over time. Nevertheless, when the matrix is not matched for both recordings, the OH concentrations are different, and the structures can occur as additional absorptions or negative absorptions (like 'emissions'), depending on their ratio. These effects are clearly visible, e.g. in the overview spectra of Figures 7.17, 7.21, 7.31, 7.42 and 7.52. Moreover, in the case of LS AAS investigations close to the LOD, even the statistical fluctuations of the flame itself are high enough to increase the background noise levels or to result in small BC errors due to the fact that the D_2 correction is not wavelength selective. Further details and the unique least-squares BC procedure offered by HR-CS AAS to overcome such problems are discussed in Sections 5.2 and 8.1.1.

Again, weak NO bands as well as unknown molecule structures, as complex as those obtained when analyzing sulfuric acid (see Section 7.2.7), are present in the wavelength region below 240 nm.

The standard measurement scheme was used for signal registration, consisting of 2×50 individual recordings with an illumination time of 10 s. In the case of sample measurements the nitrous oxide/acetylene flame was turned on, while for blank signal registration the flame was turned off. No additional solution was aspirated by the nebulizer to get the pure spectrum of the flame gases.

7.2.14 PO

The excitation of the PO molecule, as shown in the level scheme of Figure 7.41, is more complex than that of the other molecules mentioned before. In total, there exist three different electronic transitions responsible for band systems in the wavelength region of interest. An overview spectrum of the PO structures is depicted in Figure 7.42. Starting from the $X\,^2\Pi$ ground state, the dominating excitation results from a transition to the $A\,^2\Sigma^+$ state located between 220 nm and 264 nm. The corresponding systems can be further subdivided into transitions according to their different change in vibrational excitation ($\Delta v = 0, \pm 1, \pm 2, \ldots$). The decreasing subsystems are shifted to the red spectral range whereas the increasing ones are shifted to the blue region. Spectrum details of the corresponding bands are shown in Figures 7.43–7.47.

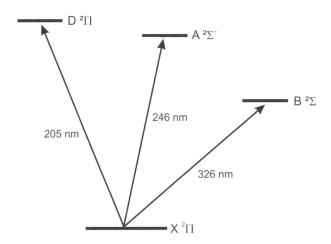

Figure 7.41: Electronic level scheme of the PO molecule and corresponding center wavelengths for transitions without change in the vibrational excitation ($\Delta v = 0$)

At the short wavelength end weaker PO structures are visible between 200 nm and 212 nm (refer to Figures 7.42 and 7.48). They are caused by a $\Delta v = 0$ sequence of the electronic transition to the $D\,^2\Pi$ state. The third electronic transition to the $B\,^2\Sigma$ state generates molecule bands in the wavelength region around 326 nm. It is not as fine-structured as the others and mainly results in the three sharp band heads at 324.62 nm, 325.53 nm and 327.05 nm (see Figure 7.49). The corresponding characteristics of all PO bands are compiled in Table 7.13.

7.2 Individual Overview Spectra

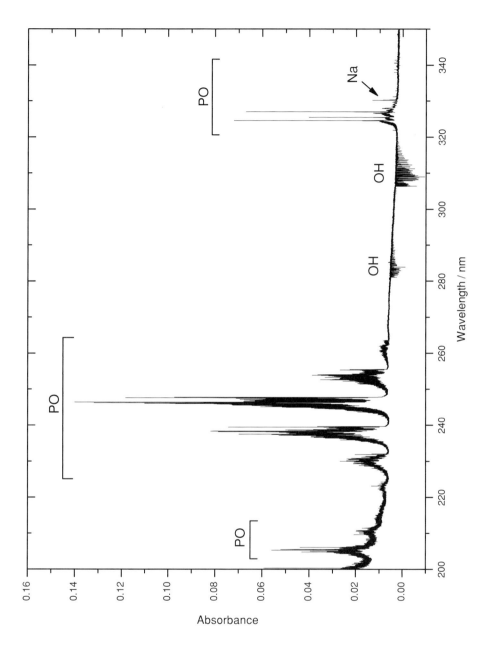

Figure 7.42: Overview spectrum of P in an air / acetylene flame with bands of PO (sample: 5 % v/v H_3PO_4; reference: deionized H_2O)

7. Electron Excitation Spectra of Diatomic Molecules

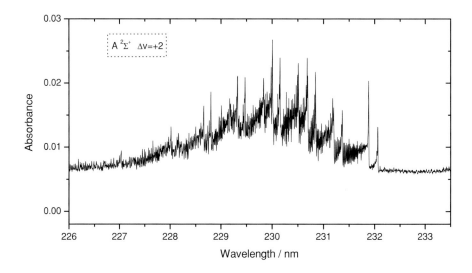

Figure 7.43: Spectrum detail of an air/acetylene flame showing the $\Delta v = +2$ sequence of the PO band system around 230 nm; electronic transition $X\,^2\Pi \rightarrow A\,^2\Sigma^+$ (sample: 5 % v/v H_3PO_4; reference: deionized H_2O)

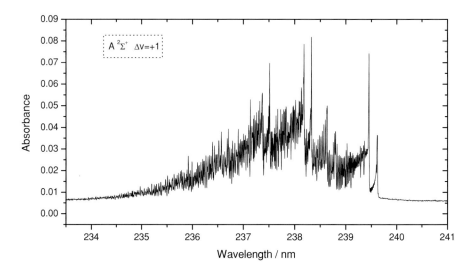

Figure 7.44: Spectrum detail of an air/acetylene flame showing the $\Delta v = +1$ sequence of the PO band system around 238 nm; electronic transition $X\,^2\Pi \rightarrow A\,^2\Sigma^+$ (sample: 5 % v/v H_3PO_4; reference: deionized H_2O)

7.2 Individual Overview Spectra

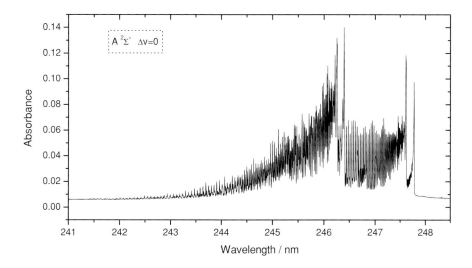

Figure 7.45: Spectrum detail of an air/acetylene flame showing the $\Delta v = 0$ sequence of the PO band system around 246 nm; electronic transition $X\,^2\Pi \rightarrow A\,^2\Sigma^+$ (sample: 5 % v/v H_3PO_4; reference: deionized H_2O)

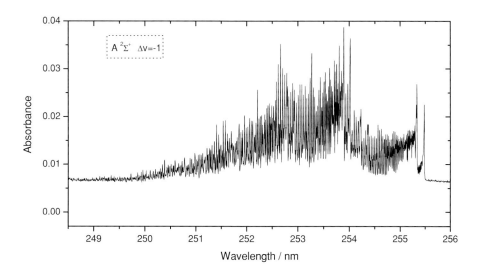

Figure 7.46: Spectrum detail of an air/acetylene flame showing the $\Delta v = -1$ sequence of the PO band system around 253 nm; electronic transition $X\,^2\Pi \rightarrow A\,^2\Sigma^+$ (sample: 5 % v/v H_3PO_4; reference: deionized H_2O)

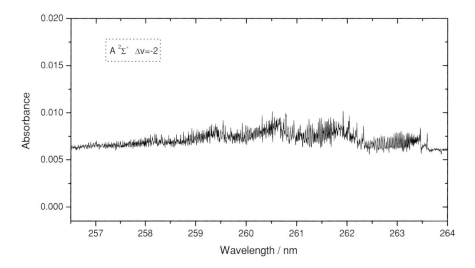

Figure 7.47: Spectrum detail of an air/acetylene flame showing the $\Delta v = -2$ sequence of the PO band system around 261 nm; electronic transition $X\,^2\Pi \rightarrow A\,^2\Sigma^+$ (sample: 5 % v/v H_3PO_4; reference: deionized H_2O)

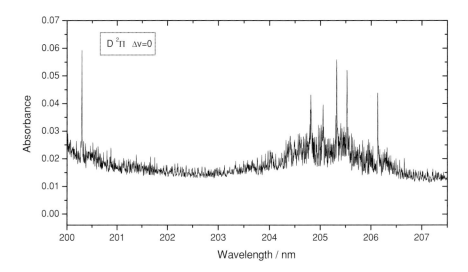

Figure 7.48: Spectrum detail of an air/acetylene flame showing the $\Delta v = 0$ sequence of the PO band system around 205 nm; electronic transition $X\,^2\Pi \rightarrow D\,^2\Pi$ (sample: 5 % v/v H_3PO_4; reference: deionized H_2O)

7.2 Individual Overview Spectra

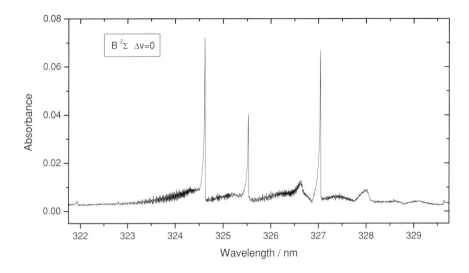

Figure 7.49: Spectrum detail of an air/acetylene flame showing the $\Delta v = 0$ sequence of the PO band system around 326 nm; electronic transition $X\,^2\Pi \rightarrow B\,^2\Sigma$ (sample: 5 % v/v H_3PO_4; reference: deionized H_2O)

Table 7.13: Characteristics of the PO bands

Wavelength / nm	Relative intensity	Upper electronic energy / eV	Electronic transition	Vibrational transition
205.31	1000	?	$X\,^2\Pi \rightarrow D\,^2\Pi$	$0 \rightarrow 0$
206.13	1000	?	$X\,^2\Pi \rightarrow D\,^2\Pi$	$0 \rightarrow 0$
210.59	800	?	$X\,^2\Pi \rightarrow D\,^2\Pi$	$1 \rightarrow 0$
211.45	800	?	$X\,^2\Pi \rightarrow D\,^2\Pi$	$1 \rightarrow 0$
216.09	600	?	$X\,^2\Pi \rightarrow D\,^2\Pi$	$2 \rightarrow 0$
216.99	600	?	$X\,^2\Pi \rightarrow D\,^2\Pi$	$2 \rightarrow 0$
228.82	200	5.01	$X\,^2\Pi \rightarrow A\,^2\Sigma^+$	$3 \rightarrow 5$
229.49	300	5.01	$X\,^2\Pi \rightarrow A\,^2\Sigma^+$	$2 \rightarrow 4$
230.04	300	5.01	$X\,^2\Pi \rightarrow A\,^2\Sigma^+$	$3 \rightarrow 5$
230.17	400	5.01	$X\,^2\Pi \rightarrow A\,^2\Sigma^+$	$1 \rightarrow 3$
230.69	400	5.01	$X\,^2\Pi \rightarrow A\,^2\Sigma^+$	$2 \rightarrow 4$
231.37	400	5.01	$X\,^2\Pi \rightarrow A\,^2\Sigma^+$	$1 \rightarrow 3$

Table 7.13 (continued)

Wavelength / nm	Relative intensity	Upper electronic energy / eV	Electronic transition	Vibrational transition
232.06	300	5.01	$X\,^2\Pi \to A\,^2\Sigma^+$	$0 \to 2$
236.73	200	5.01	$X\,^2\Pi \to A\,^2\Sigma^+$	$2 \to 3$
237.52	700	5.01	$X\,^2\Pi \to A\,^2\Sigma^+$	$1 \to 2$
237.99	200	5.01	$X\,^2\Pi \to A\,^2\Sigma^+$	$2 \to 3$
238.35	800	5.01	$X\,^2\Pi \to A\,^2\Sigma^+$	$0 \to 1$
238.79	700	5.01	$X\,^2\Pi \to A\,^2\Sigma^+$	$1 \to 2$
239.63	800	5.01	$X\,^2\Pi \to A\,^2\Sigma^+$	$0 \to 1$
245.46	600	5.01	$X\,^2\Pi \to A\,^2\Sigma^+$	$1 \to 1$
245.90	200	5.01	$X\,^2\Pi \to A\,^2\Sigma^+$	$2 \to 2$
246.42	1000	5.01	$X\,^2\Pi \to A\,^2\Sigma^+$	$0 \to 0$
246.83	600	5.01	$X\,^2\Pi \to A\,^2\Sigma^+$	$1 \to 1$
247.79	1000	5.01	$X\,^2\Pi \to A\,^2\Sigma^+$	$0 \to 0$
251.87	500	5.01	$X\,^2\Pi \to A\,^2\Sigma^+$	$3 \to 2$
252.94	700	5.01	$X\,^2\Pi \to A\,^2\Sigma^+$	$2 \to 1$
254.04	1000	5.01	$X\,^2\Pi \to A\,^2\Sigma^+$	$1 \to 0$
254.39	700	5.01	$X\,^2\Pi \to A\,^2\Sigma^+$	$2 \to 1$
255.50	1000	5.01	$X\,^2\Pi \to A\,^2\Sigma^+$	$1 \to 0$
260.80	700	5.01	$X\,^2\Pi \to A\,^2\Sigma^+$	$3 \to 1$
261.07	200	5.01	$X\,^2\Pi \to A\,^2\Sigma^+$	$4 \to 2$
262.05	800	5.01	$X\,^2\Pi \to A\,^2\Sigma^+$	$2 \to 0$
262.34	700	5.01	$X\,^2\Pi \to A\,^2\Sigma^+$	$3 \to 1$
263.63	700	5.01	$X\,^2\Pi \to A\,^2\Sigma^+$	$2 \to 0$
324.62	1000	3.82	$X\,^2\Pi \to B\,^2\Sigma$	$0 \to 0$
325.53	800	3.82	$X\,^2\Pi \to B\,^2\Sigma$	$1 \to 1$
327.05	900	3.82	$X\,^2\Pi \to B\,^2\Sigma$	$0 \to 0$
330.28	500	3.82	$X\,^2\Pi \to B\,^2\Sigma$	$5 \to 5$
331.18	500	3.82	$X\,^2\Pi \to B\,^2\Sigma$	$4 \to 4$
332.09	700	3.82	$X\,^2\Pi \to B\,^2\Sigma$	$6 \to 6$
332.82	700	3.82	$X\,^2\Pi \to B\,^2\Sigma$	$5 \to 5$
337.98	600	3.82	$X\,^2\Pi \to B\,^2\Sigma$	$6 \to 6$
338.76	300	3.82	$X\,^2\Pi \to B\,^2\Sigma$	$6 \to 6$
338.79	600	3.82	$X\,^2\Pi \to B\,^2\Sigma$	$6 \to 6$

Signal registration was accomplished using the standard measurement scheme, where 2×50 spectra with an individual illumination time of 10 s were recorded. A 5 % v/v H_3PO_4 solution was used as the sample and deionized water as the reference.

7.2.15 SH

It has already been mentioned in Section 7.2.7 that when using sulfuric acid (H_2SO_4) as sample solution, apart from CS structures, SH molecule bands are also generated in the conventional air/acetylene flame. They are located in the middle wavelength region around 326 nm resulting from a $\Delta v = 0$ sequence of the electronic transition between the $X\,^2\Pi$ and the $A\,^2\Sigma^+$ state. The detailed spectrum of the SH structure is shown in Figure 7.50 and the corresponding characteristics of the SH bands are summarized in Table 7.14.

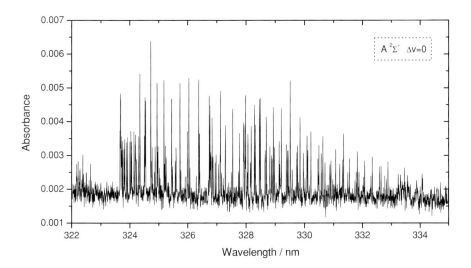

Figure 7.50: Spectrum detail of an air/acetylene flame showing the $\Delta v = 0$ sequence of the SH band system around 326 nm; electronic transition $X\,^2\Pi \to A\,^2\Sigma^+$ (sample: 5 % v/v H_2SO_4; reference: deionized H_2O)

The same standard measurement scheme, consisting of 2×50 recordings with an illumination time of 10 s each, was used for signal registration as applied in Section 7.2.7 to obtain the CS molecule bands. Again, a 5 % v/v H_2SO_4 solution was used as the sample and deionized water as the reference.

7. Electron Excitation Spectra of Diatomic Molecules

Table 7.14: Characteristics of the SH bands

Wavelength / nm	Relative intensity	Upper electronic energy / eV	Electronic transition	Vibrational transition
323.66	800	?	$X\,^2\Pi \rightarrow A\,^2\Sigma^+$	$0 \rightarrow 0$
324.07	700	?	$X\,^2\Pi \rightarrow A\,^2\Sigma^+$	$0 \rightarrow 0$
327.91	400	?	$X\,^2\Pi \rightarrow A\,^2\Sigma^+$	$0 \rightarrow 0$

7.2.16 SiO

The analysis of the HISS-1 Sediment reference material (NRC, Canada) in an air/acetylene flame yielded pronounced SiO molecule bands having a fine structure. Because of the complex matrix, a lot of atomic absorption lines are present, as well as structures of the NO molecule. All SiO structures result from the same electronic transition between the $X\,^1\Sigma^+$ and the $A\,^1\Pi$ state. According to their change in vibrational excitation, they can be subdivided into sequences starting with $\Delta v = +5$ at about 216 nm and going down to $\Delta v = -3$ at about 258 nm (refer to Figure 7.51). In total, they stretch over a wavelength region from 210 nm to 260 nm, as can be seen in Figure 7.52. The corresponding spectra are depicted in detail in Figures 7.53–7.61 and the characteristics of the bands are compiled in Table 7.15.

For signal registration the standard measurement scheme was used, consisting of 2×20 recordings with an illumination time of 10 s. The HISS-1 Sediment reference material, prepared in concentrated HNO_3 + 2 % v/v HF and subsequently dissolved in 2 % v/v HNO_3, was used as the sample and deionized water as the reference.

Table 7.15: Characteristics of the SiO bands

Wavelength / nm	Relative intensity	Upper electronic energy / eV	Electronic transition	Vibrational transition
216.03	?	5.31	$X\,^1\Sigma^+ \rightarrow A\,^1\Pi$	$1 \rightarrow 6$
217.66	100	5.31	$X\,^1\Sigma^+ \rightarrow A\,^1\Pi$	$0 \rightarrow 4$
219.74	?	5.31	$X\,^1\Sigma^+ \rightarrow A\,^1\Pi$	$1 \rightarrow 5$
221.54	200	5.31	$X\,^1\Sigma^+ \rightarrow A\,^1\Pi$	$0 \rightarrow 3$
223.63	200	5.31	$X\,^1\Sigma^+ \rightarrow A\,^1\Pi$	$1 \rightarrow 4$

7.2 Individual Overview Spectra

Table 7.15 (continued)

Wavelength / nm	Relative intensity	Upper electronic energy / eV	Electronic transition	Vibrational transition
225.59	400	5.31	$X\,^1\Sigma^+ \to A\,^1\Pi$	$0 \to 2$
227.72	100	5.31	$X\,^1\Sigma^+ \to A\,^1\Pi$	$1 \to 3$
229.89	600	5.31	$X\,^1\Sigma^+ \to A\,^1\Pi$	$0, 2 \to 1, 4$
234.24	100	5.31	$X\,^1\Sigma^+ \to A\,^1\Pi$	$2 \to 3$
234.43	500	5.31	$X\,^1\Sigma^+ \to A\,^1\Pi$	$0 \to 0$
236.57	600	5.31	$X\,^1\Sigma^+ \to A\,^1\Pi$	$1 \to 1$
238.79	500	5.31	$X\,^1\Sigma^+ \to A\,^1\Pi$	$2 \to 2$
241.38	700	5.31	$X\,^1\Sigma^+ \to A\,^1\Pi$	$1 \to 0$
243.63	300	5.31	$X\,^1\Sigma^+ \to A\,^1\Pi$	$2 \to 1$
248.68	600	5.31	$X\,^1\Sigma^+ \to A\,^1\Pi$	$2 \to 0$
250.99	400	5.31	$X\,^1\Sigma^+ \to A\,^1\Pi$	$3 \to 1$
256.38	500	5.31	$X\,^1\Sigma^+ \to A\,^1\Pi$	$3 \to 0$

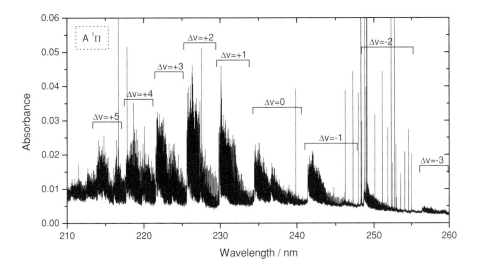

Figure 7.51: Enlarged overview spectrum of the SiO band system in an air / acetylene flame around 230 nm; electronic transition $X\,^1\Sigma^+ \to A\,^1\Pi$ (sample: 2 % m/v HISS-1 Sediment reference material in 2 % v/v HNO_3; reference: deionized H_2O)

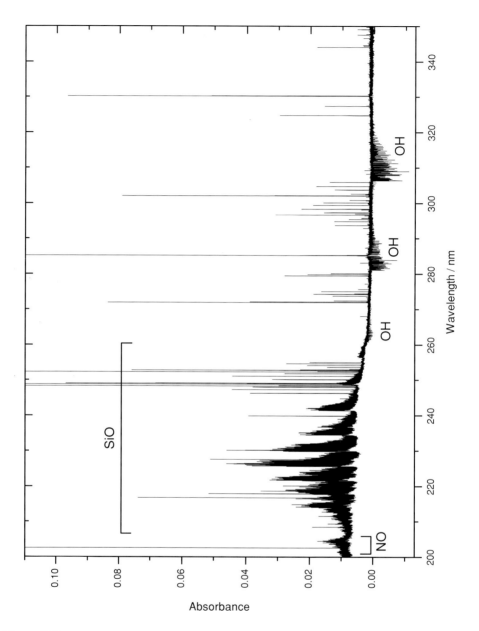

Figure 7.52: Overview spectrum of Si in an air / acetylene flame with bands of SiO and NO (sample: 2 % m/v HISS-1 Sediment reference material in 2 % v/v HNO_3; reference: deionized H_2O)

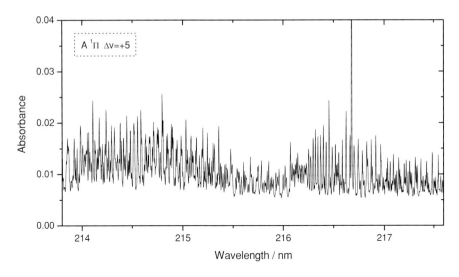

Figure 7.53: Spectrum detail of an air/acetylene flame showing the $\Delta v = +5$ sequence of the SiO band system around 216 nm; electronic transition $X\,^1\Sigma^+ \to A\,^1\Pi$ (sample: 2 % m/v HISS-1 Sediment reference material in 2 % v/v HNO_3; reference: deionized H_2O)

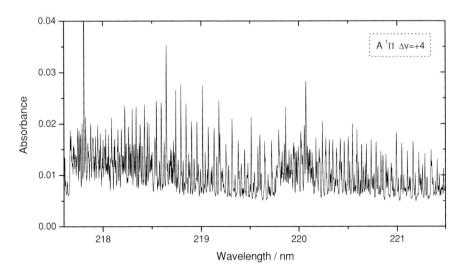

Figure 7.54: Spectrum detail of an air/acetylene flame showing the $\Delta v = +4$ sequence of the SiO band system around 219 nm; electronic transition $X\,^1\Sigma^+ \to A\,^1\Pi$ (sample: 2 % m/v HISS-1 Sediment reference material in 2 % v/v HNO_3; reference: deionized H_2O)

7. Electron Excitation Spectra of Diatomic Molecules

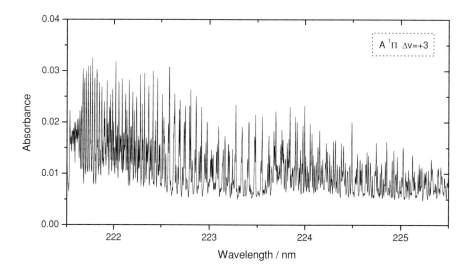

Figure 7.55: Spectrum detail of an air / acetylene flame showing the $\Delta v = +3$ sequence of the SiO band system around 223 nm; electronic transition $X\,^1\Sigma^+ \rightarrow A\,^1\Pi$ (sample: 2 % m/v HISS-1 Sediment reference material in 2 % v/v HNO_3; reference: deionized H_2O)

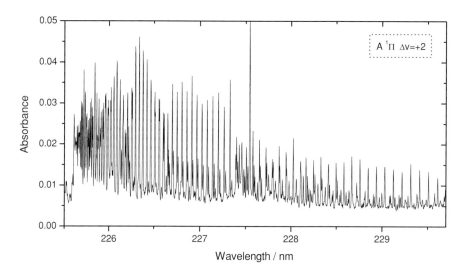

Figure 7.56: Spectrum detail of an air / acetylene flame showing the $\Delta v = +2$ sequence of the SiO band system around 227 nm; electronic transition $X\,^1\Sigma^+ \rightarrow A\,^1\Pi$ (sample: 2 % m/v HISS-1 Sediment reference material in 2 % v/v HNO_3; reference: deionized H_2O)

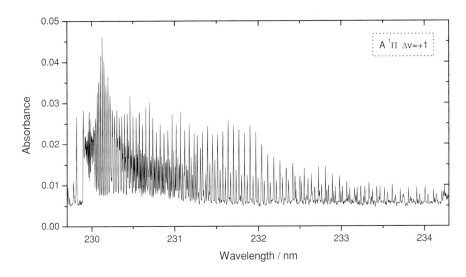

Figure 7.57: Spectrum detail of an air/acetylene flame showing the $\Delta v = +1$ sequence of the SiO band system around 232 nm; electronic transition $X\,^1\Sigma^+ \rightarrow A\,^1\Pi$ (sample: 2 % m/v HISS-1 Sediment reference material in 2 % v/v HNO_3; reference: deionized H_2O)

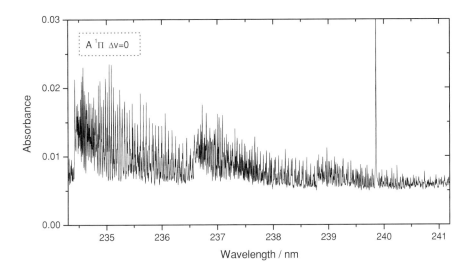

Figure 7.58: Spectrum detail of an air/acetylene flame showing the $\Delta v = 0$ sequence of the SiO band system around 237 nm; electronic transition $X\,^1\Sigma^+ \rightarrow A\,^1\Pi$ (sample: 2 % m/v HISS-1 Sediment reference material in 2 % v/v HNO_3; reference: deionized H_2O)

7. Electron Excitation Spectra of Diatomic Molecules

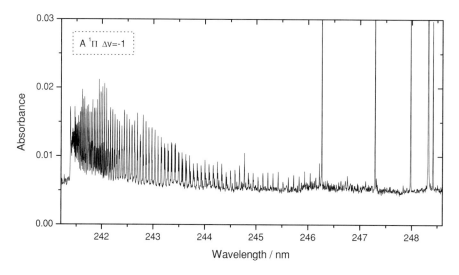

Figure 7.59: Spectrum detail of an air/acetylene flame showing the $\Delta v = -1$ sequence of the SiO band system around 244 nm; electronic transition $X\,^1\Sigma^+ \rightarrow A\,^1\Pi$ (sample: 2 % m/v HISS-1 Sediment reference material in 2 % v/v HNO_3; reference: deionized H_2O)

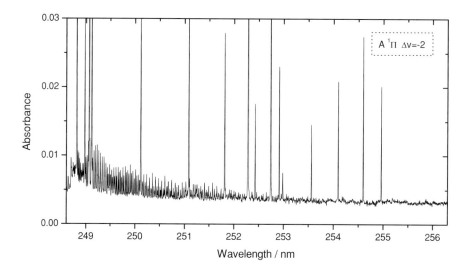

Figure 7.60: Spectrum detail of an air/acetylene flame showing the $\Delta v = -2$ sequence of the SiO band system around 250 nm; electronic transition $X\,^1\Sigma^+ \rightarrow A\,^1\Pi$ (sample: 2 % m/v HISS-1 Sediment reference material in 2 % v/v HNO_3; reference: deionized H_2O)

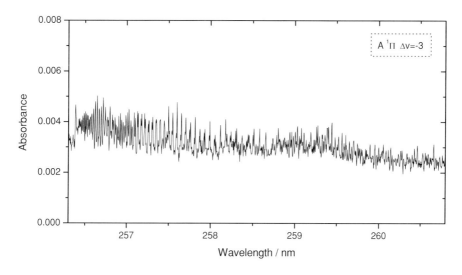

Figure 7.61: Spectrum detail of an air/acetylene flame showing the $\Delta v = -3$ sequence of the SiO band system around 258 nm; electronic transition $X\,^1\Sigma^+ \to A\,^1\Pi$ (sample: 2 % m/v HISS-1 Sediment reference material in 2 % v/v HNO_3; reference: deionized H_2O)

7.2.17 SnO

The overview spectrum of a 10 g/L Sn solution, prepared from $SnCl_4$ dissolved in 5 % v/v HCl, which is depicted in Figure 7.62, shows a multitude of Sn atomic absorption lines, as well as two groups of molecule bands. They result either from the electronic transition starting at the $X\,^1\Sigma^+$ ground state and going to the $E\,^1\Sigma^+$ state or from the one which connects to the $D\,^1\Pi$ state. Again, both are fine-structured and can be further divided into systems with the same change in vibrational excitation.

The first system, shown in Figure 7.63, covers the wavelength region from about 230 nm to 290 nm while the latter one can be found in the region between 300 nm and 370 nm (refer to Figure 7.66). Their individual structure is, of course, too complex to be depicted completely here with all the information included in the original recordings, therefore, only two examples with the strongest bands for each electronic transition are given. The corresponding spectra are shown in Figures 7.64 and 7.65 for the $E\,^1\Sigma^+$ state as well as in Figures 7.67 and 7.68 for the $D\,^1\Pi$ state. A complete list of the characteristics of the SnO bands is compiled in Table 7.16.

Once more, the standard measurement scheme was used for signal registration, consisting of 2×50 recordings with an individual illumination time of 17 s and deionized water was utilized as reference sample.

7. Electron Excitation Spectra of Diatomic Molecules

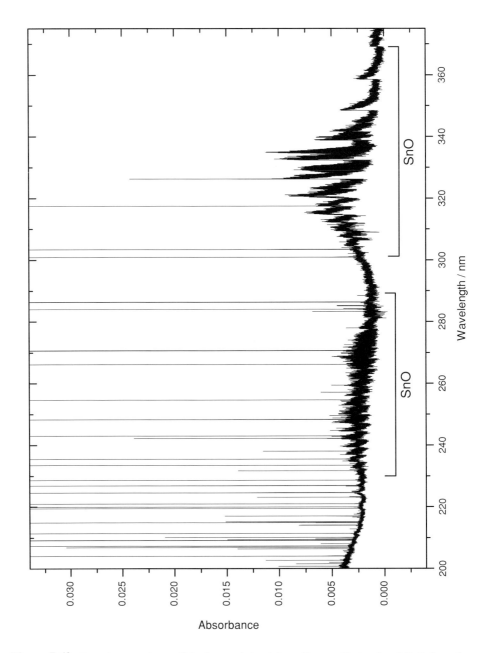

Figure 7.62: Overview spectrum of Sn in an air/acetylene flame with bands of SnO (sample: 10 g/L Sn as $SnCl_4$ dissolved in 5 % v/v HCl; reference: deionized H_2O)

7.2 Individual Overview Spectra

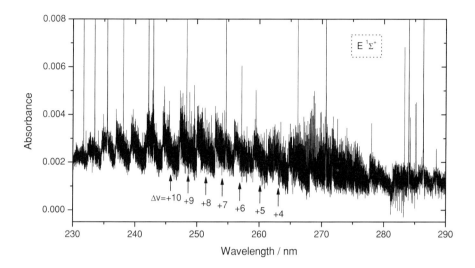

Figure 7.63: Enlarged overview spectrum of the SnO band system in an air/acetylene flame around 260 nm; electronic transition $X\,^1\Sigma^+ \rightarrow E\,^1\Sigma^+$ (sample: 10 g/L Sn as SnCl$_4$ dissolved in 5 % v/v HCl; reference: deionized H$_2$O)

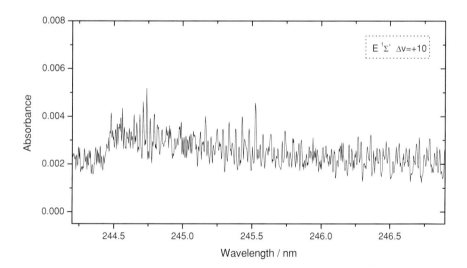

Figure 7.64: Spectrum detail of an air/acetylene flame showing the $\Delta v = +10$ sequence of the SnO band system around 245 nm; electronic transition $X\,^1\Sigma^+ \rightarrow E\,^1\Sigma^+$ (sample: 10 g/L Sn as SnCl$_4$ dissolved in 5 % v/v HCl; reference: deionized H$_2$O)

7. Electron Excitation Spectra of Diatomic Molecules

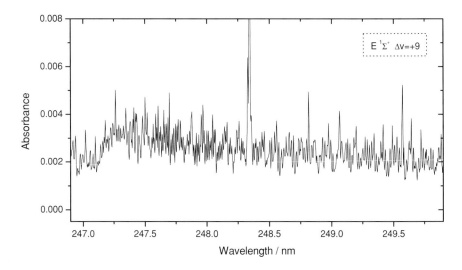

Figure 7.65: Spectrum detail of an air/acetylene flame showing the $\Delta v = +9$ sequence of the SnO band system around 248 nm; electronic transition $X\,^1\Sigma^+ \to E\,^1\Sigma^+$ (sample: 10 g/L Sn as $SnCl_4$ dissolved in 5 % v/v HCl; reference: deionized H_2O)

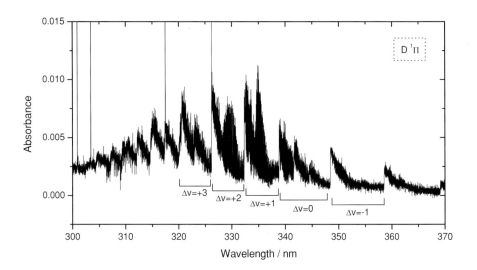

Figure 7.66: Enlarged overview spectrum of the SnO band system in an air/acetylene flame around 335 nm; electronic transition $X\,^1\Sigma^+ \to D\,^1\Pi$ (sample: 10 g/L Sn as $SnCl_4$ dissolved in 5 % v/v HCl; reference: deionized H_2O)

7.2 Individual Overview Spectra

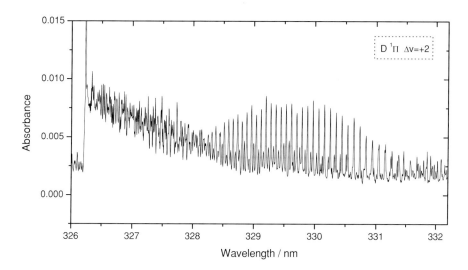

Figure 7.67: Spectrum detail of an air / acetylene flame showing the $\Delta v = +2$ sequence of the SnO band system around 329 nm; electronic transition $X\ ^1\Sigma^+ \rightarrow D\ ^1\Pi$ (sample: 10 g/L Sn as SnCl$_4$ dissolved in 5 % v/v HCl; reference: deionized H$_2$O)

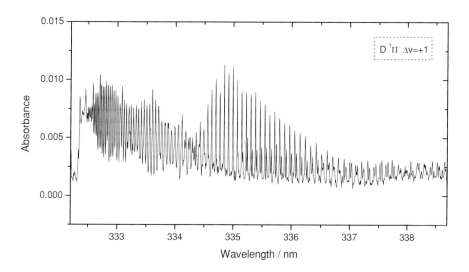

Figure 7.68: Spectrum detail of an air / acetylene flame showing the $\Delta v = +1$ sequence of the SnO band system around 335 nm; electronic transition $X\ ^1\Sigma^+ \rightarrow D\ ^1\Pi$ (sample: 10 g/L Sn as SnCl$_4$ dissolved in 5 % v/v HCl; reference: deionized H$_2$O)

7. Electron Excitation Spectra of Diatomic Molecules

Table 7.16: Characteristics of the SnO bands

Wavelength / nm	Relative intensity	Upper electronic energy / eV	Electronic transition	Vibrational transition
244.50	100	4.50	$X^1\Sigma^+ \to E^1\Sigma^+$	$0 \to 10$
247.23	200	4.50	$X^1\Sigma^+ \to E^1\Sigma^+$	$0 \to 9$
250.04	200	4.50	$X^1\Sigma^+ \to E^1\Sigma^+$	$0 \to 8$
252.99	300	4.50	$X^1\Sigma^+ \to E^1\Sigma^+$	$0 \to 7$
256.00	300	4.50	$X^1\Sigma^+ \to E^1\Sigma^+$	$0 \to 6$
264.69	600	4.50	$X^1\Sigma^+ \to E^1\Sigma^+$	$1 \to 5$
265.81	600	4.50	$X^1\Sigma^+ \to E^1\Sigma^+$	$0 \to 3$
268.08	600	4.50	$X^1\Sigma^+ \to E^1\Sigma^+$	$1 \to 4$
271.69	700	4.50	$X^1\Sigma^+ \to E^1\Sigma^+$	$1 \to 3$
274.01	600	4.50	$X^1\Sigma^+ \to E^1\Sigma^+$	$2 \to 4$
281.48	500	4.50	$X^1\Sigma^+ \to E^1\Sigma^+$	$2 \to 2$
292.17	200	4.50	$X^1\Sigma^+ \to E^1\Sigma^+$	$3 \to 1$
294.77	?	4.50	$X^1\Sigma^+ \to E^1\Sigma^+$	$4 \to 2$
299.04	?	4.50	$X^1\Sigma^+ \to E^1\Sigma^+$	$4 \to 1$
304.36	?	4.50	$X^1\Sigma^+ \to E^1\Sigma^+$	$6 \to 3$
320.58	500	?	$X^1\Sigma^+ \to D^1\Pi$	$0 \to 3$
326.24	750	?	$X^1\Sigma^+ \to D^1\Pi$	$0 \to 2$
332.34	870	?	$X^1\Sigma^+ \to D^1\Pi$	$0 \to 1$
338.83	750	?	$X^1\Sigma^+ \to D^1\Pi$	$0 \to 0$
341.58	620	?	$X^1\Sigma^+ \to D^1\Pi$	$1 \to 1$
348.45	1000	?	$X^1\Sigma^+ \to D^1\Pi$	$1 \to 0$
358.54	870	?	$X^1\Sigma^+ \to D^1\Pi$	$2 \to 0$
369.14	620	?	$X^1\Sigma^+ \to D^1\Pi$	$3 \to 0$

8. Specific Applications

8.1 Flame Measurements

8.1.1 Molecular Background in Flame AAS

In Chapter 7 several electron excitation spectra of diatomic molecules that have been recorded in a flame were shown. Some of these spectra are only observed in the presence of specific matrices or acids, such as those of the PO and CS molecules, whereas others, particularly the spectrum of OH, is present in any spectroscopic flame, even when no sample is introduced. The concentrations of radicals such as O, OH, CN and H in a flame are determined by the reactions between the natural components of the flame. The influence of any sample constituents on these components is negligible, since every reaction that leads to a reduction or an increase in the concentration of such a species is immediately counteracted by infinitesimal shifts in the equilibria of the main components. This means that these molecular absorption structures in a flame should remain constant, and be eliminated in the calibration process. However, this is not exactly the case, as will be shown in the following examples.

Firstly, when the spectrum of OH shown in Figure 7.37 (refer to Section 7.2.13) is compared with that obtained for dilute phosphoric or sulfuric acid (Figures 7.21 and 7.42, respectively), it becomes obvious that the OH spectrum appears as a 'negative absorbance' in these spectra. This artifact is caused by the fact that, although sample constituents cannot influence the equilibria in a flame, matrix constituents when present at high concentration can. This means that matrix-matched blank and calibration solutions have to be used whenever an atomic absorption line coincides with a molecular band in order to avoid measurement errors. While the use of matrix-matched standards is the only solution in LS AAS, unless an alternate absorption line is chosen that is not affected by any molecular absorption, HR-CS AAS offers the very efficient possibility of least-squares background correction (BC) to completely eliminate this problem, as will be shown in the following. Additional details about this unique BC procedure are given in Section 5.2.2.

8. Specific Applications

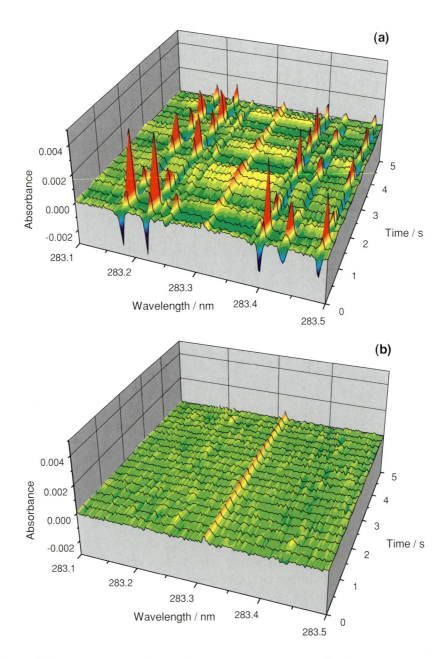

Figure 8.1: Influence of the OH molecular absorption in the air/acetylene flame on the determination of lead in the vicinity of the 283.306 nm line; aqueous standard 100 µg/L Pb; (a) without correction; (b) with least-squares BC

Secondly, even in the absence of any matrix, or when the matrix is absolutely identical in all solutions for measurement, the concentration of molecules such as OH is not strictly constant over time, but subject to some short-term fluctuations that result in kind of a flicker-noise, as shown in Figure 8.1 (a) in the vicinity of the lead absorption line at 283.306 nm. These fluctuations obviously result in positive as well as negative 'peaks' after the usual baseline correction. As least-squares BC corrects each individual spectrum separately, it can handle any kind of positive or negative deflection, eliminating completely the noise produced by the molecular absorption, as shown in Figure 8.1 (b).

As in the case of lead the atomic line does not overlap with any molecular structure, no improvement in the LOD is obtained by the use of least-squares BC. However, it should be mentioned that additional noise, and even measurement errors, could be introduced in LS AAS when deuterium background correction is used in this case.

A significantly different situation is faced in the case of the bismuth absorption line at 306.772 nm, which directly overlaps with a molecular absorption band of OH, as shown in Figure 8.2 (a). In this case least-squares BC not only corrects for the noise in the vicinity of the atomic line, as shown in Figure 8.2 (b), but also improves significantly the SNR of the measurement, and hence the LOD that can be obtained at that absorption line.

8.1.2 Drinking Water Analysis

Schlemmer et al. [125] have carried out a direct comparison between HR-CS AAS and LS AAS for the determination of a number of trace elements in drinking water, a determination that was not expected to present any difficulties, but which is carried out on a large scale in many routine laboratories.

The elements investigated were cadmium, chromium, lead and zinc, and there was no statistically significant difference between the results on a 95 % confidence level. For cadmium at 228.802 nm a 2 – 3 times better detection limit was obtained using HR-CS AAS, and the precision was also significantly better, particularly when compared with LS AAS using a deuterium background corrector, where the difference was almost a factor of 5. Significantly better LOD were also obtained by HR-CS AAS for lead at both analytical lines, with a clear advantage for the 217.001 nm line. For zinc at 213.856 nm the LOD were also better by about a factor of two using HR-CS AAS.

The most pronounced improvement of at least a factor of five for both LOD and precision was obtained for the determination of chromium, demonstrating one more time the superiority of HR-CS AAS, particularly for 'difficult' elements, even in simple matrices such as water.

8. Specific Applications

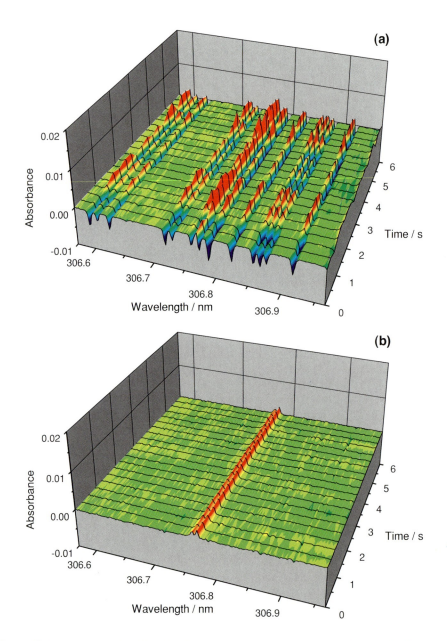

Figure 8.2: Influence of the OH molecular absorption in the air / acetylene flame on the determination of bismuth in the vicinity of the 306.772 nm line; aqueous standard 500 μg/L Bi; (a) without correction; (b) with least-squares BC

8.1.3 Sodium and Potassium in Animal Food and Pharmaceutical Products

Schlemmer et al. [125] investigated the determination of medium and high concentrations of sodium and potassium in animal food and pharmaceutical products, comparing LS AAS, flame OES and HR-CS AAS. For the determination of sodium at medium levels in pharmaceutical products, the linear working range for HR-CS AAS was somewhat greater than with LS AAS, and about three times better than in flame OES, with similar LOD for the three techniques. For the determination of high concentrations in pharmaceutical products, with the burner head rotated 90° in order to reduce the sensitivity, very similar results were obtained with HR-CS AAS and LS AAS, and the linear working range of both AAS techniques was clearly better than that of flame OES.

For potassium in animal food, i.e. for medium concentrations, the LOD of HR-CS AAS was essentially the same as for flame OES, and a factor of two better than that of LS AAS, but, more importantly, the linear working range was almost one order of magnitude greater for HR-CS AAS than with the other two techniques. Very similar results have also been obtained for the determination of high concentrations of potassium in pharmaceutical products with the burner head rotated 90°, with a linear working range about one order of magnitude greater than with the other techniques, and also an improved precision of the results.

8.1.4 Determination of Zinc in Iron and Steel

There is a direct line interference of the 213.859 nm iron line on the 213.856 nm zinc line, as shown in Figure 8.3 (a). Even HR-CS AAS cannot resolve this coincidence of two lines that are only 3 pm apart. However, fortunately there is yet another iron line within the spectral window at 213.970 nm, which can be used as a reference line for least-squares BC, as the ratio between the intensities of the two absorption lines remains constant. The efficiency of correction by using the Zn-free iron matrix (Figure 8.3 (b)) as reference, which only leaves the zinc line within the spectral window, is shown in Figure 8.3 (c).

Obviously, the degree of interference that this coincidence causes depends greatly on the concentration of zinc in the sample to be investigated, and is most pronounced for very low zinc concentrations. Schlemmer et al. [125] analyzed several certified steel samples using HR-CS AAS and LS AAS and, according to expectation, the results of the former technique, applying least-squares BC to correct for the line overlap, were in good agreement with the certified values, whereas those obtained with the latter technique exhibited a tendency towards systematically high results for zinc.

8. Specific Applications

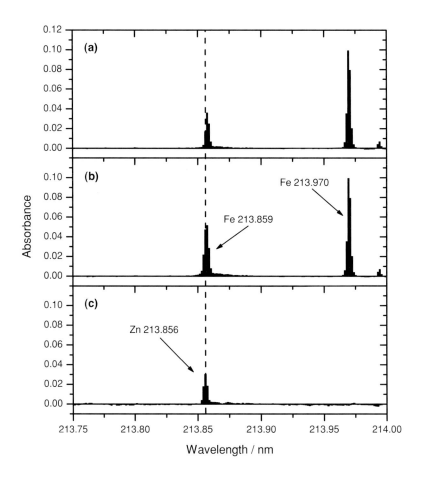

Figure 8.3: Interference of iron on the determination of zinc due to direct line overlap at 213.856 nm (dashed line); 0.3 mg/L Zn in 1000 mg/L Fe matrix; air/acetylene flame; (a) without correction; (b) Zn-free iron matrix alone; (c) with least-squares BC

8.1.5 Determination of Trace Elements in High-purity Copper

High-purity copper is one of the important materials used in electrical devices, and the content of several trace elements has to be controlled carefully, as they have a significant influence on the quality and the properties of the product. Huang et al. [67] investigated the use of HR-CS F AAS for this purpose because of the superior detection power of this technique and the improved background measurement and correction capabilities. The authors found the well-known direct line overlap between the weak copper line at 213.853 nm with the zinc resonance line at 213.856 nm [159]. They also observed some molecular absorption bands in the vicinity of the zinc line, as shown in Figure 8.4 (a), which they assigned to the NO molecule [86]. Both interferences could be removed completely using least squares BC, as shown in Figure 8.4 (b). The removal of the interference could also be confirmed by analyzing a copper CRM, where excellent agreement with the certified value was obtained, while an approximately 30 % higher value was found without correction.

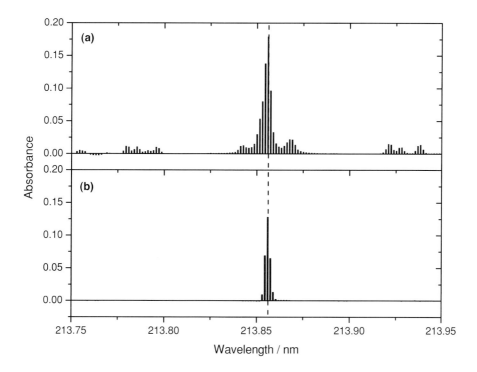

Figure 8.4: Spectral interference of copper on the determination of zinc in the vicinity of the 213.856 nm line (dashed line); (a) without correction; (b) with least-squares BC

Another relatively strong copper line at 223.008 nm in the vicinity of the bismuth line at 223.061 nm, which is shown in Figure 8.5, had no influence on the determination of that analyte, as it was well separated from the latter line. The molecular structures, however, that appeared on both sides of the copper line, had to be corrected for. In the case of cadmium, Huang et al. [67] succeeded in eliminating the molecular structures by almost doubling the oxidant flow-rate, i.e. using an extremely fuel-lean flame, as shown in Figure 8.6. Another way to eliminate the molecular absorption due to NO proposed by

8. Specific Applications

the authors was to dissolve the copper sample in hydrochloric acid and hydrogen peroxide instead of hydrochloric and nitric acids. The results obtained by the authors determining a few selected elements in two copper CRM are shown in Table 8.1.

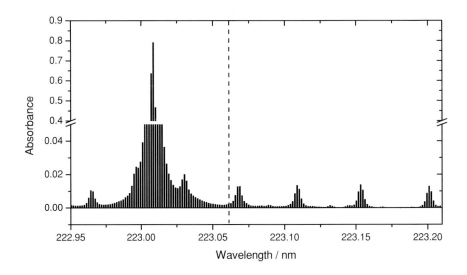

Figure 8.5: Absorbance of high copper concentrations in the vicinity of the 223.061 nm Bi line (dashed line)

Table 8.1: Results for the determination of selected trace elements in GEW 02141 copper (Shanghai Research Institute for Materials, China) and BAM 361 copper (BAM, Berlin, Germany) using HR-CS F AAS

Element	GEW 02141 Copper 99.93 %		BAM 361 Copper 99.96 %	
	found / $\mu g\ g^{-1}$	certified / $\mu g\ g^{-1}$	found / $\mu g\ g^{-1}$	certified / $\mu g\ g^{-1}$
Bi	7.27 ± 0.14	7.0	1.08 ± 0.13	1.1
Ni	38.9 ± 1.0	38.0	81.9 ± 0.5	80.0
Pb	41.7 ± 1.1	41.0	53.1 ± 0.6	48.0
Sb	22.0 ± 1.1	20.0	9.10 ± 0.24	8.0
Zn	52.9 ± 1.6	53.0	-	-

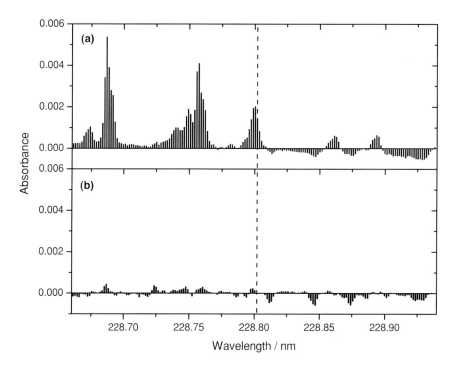

Figure 8.6: Influence of the oxidant flow-rate on the background structures in the vicinity of the 228.802 nm Cd line (dashed line) in presence of high concentrations of copper; (a) oxidant flow: 12.7 L/min; (b) oxidant flow: 24.4 L/min

8.1.6 Determination of Phosphorus via PO Molecular Absorption Lines

One of the characteristics of HR-CS AAS is that no element-specific radiation source is required, and that, in principle, absorption can be measured at any wavelength, including molecular absorption bands.

Huang et al. [68] investigated the possibility of using PO molecular absorption bands in an air/acetylene flame for analytical purposes, as described in Section 6.30. They carried out a detailed interference study, investigating the influence of a variety of acids and typical matrix elements on the PO absorption. Among the acids investigated, HCl, HF, HNO_3 and H_2SO_4, only sulfuric acid caused a slight interference at concentrations above 3 % v/v. This, however, was due to a viscosity effect, i.e. a transport interference, which could be eliminated easily by matching the acid concentration in the calibration solutions. Five new absorption lines also appeared in the vicinity of the 324.62 nm PO absorption

8. Specific Applications

band head, as shown in Figure 8.7, which were due to SH molecular absorption. Similar absorption bands were observed in the vicinity of the PO line at 327.04 nm, however, none of these bands caused any spectral interference, as they were far enough from the PO band head to be excluded from measurement.

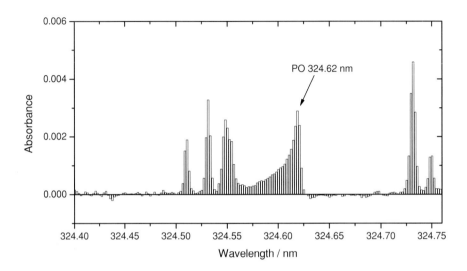

Figure 8.7: Molecular absorbance signals obtained for 200 mg/L P in the presence of 2 % v/v sulfuric acid in the vicinity of the PO band head at 324.62 nm; five SH absorption bands can be observed at both sides of the PO band head

Among the metals investigated, only iron (at 324.602 nm), as shown in Figure 8.8, and copper (at 324.754 nm) were found to cause interference at the 324.62 nm PO line, when present in high concentration, because of severe line wing broadening, as shown in Figure 8.9 for the case of the copper interference. This potential interference could easily be avoided by going to another PO line that does not show this kind of overlap.

The only really serious interference the authors found was that due to magnesium and, particularly, to calcium, as shown in Figure 8.10, which has been associated with the formation of stable magnesium and calcium phosphates in the air/acetylene flame. Huang et al. [69] found that titanium was the most efficient releasing agent, and that the addition of 1 g/L Ti removed all the interference completely, as shown in Figure 8.11. The mechanism was most likely the formation of a thermally stable compound with a Ti/Ca molar ratio of one.

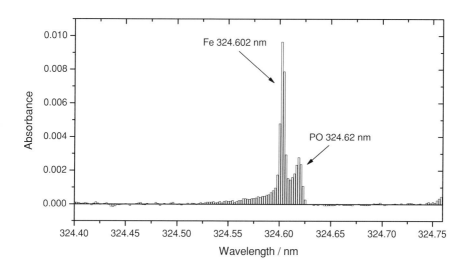

Figure 8.8: Coincidence of the iron line at 324.602 nm with the PO band head at 324.62 nm; 200 mg/L P and 1000 mg/L Fe; the iron line is eight pixels away from the PO band head

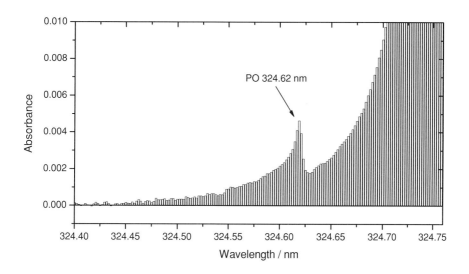

Figure 8.9: Overlap of the wing of the copper line at 324.754 nm with the PO band head at 324.62 nm; 200 mg/L P and 100 mg/L Cu

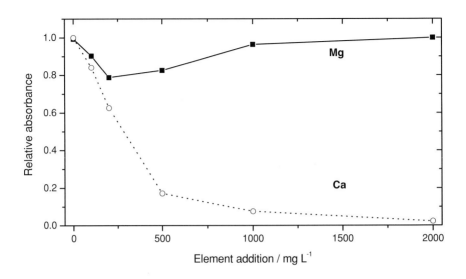

Figure 8.10: Influence of calcium and magnesium on the absorbance signal of 200 mg/L P, measured at the PO band head at 324.62 nm in an air / acetylene flame

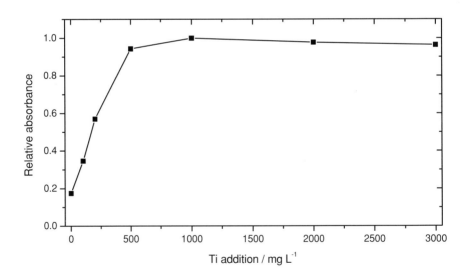

Figure 8.11: Influence of the addition of increasing concentrations of titanium as releasing agent on the interference of 500 mg/L Ca on the determination of 200 mg/L P, measured at the PO band head at 324.62 nm in an air / acetylene flame

The proposed method has been applied successfully to determine phosphorus in cast iron, in pine needles and in super phosphate fertilizer. The reported detection limit of 1.3 mg/L is almost two orders of magnitude better than that obtained by LS AAS in a nitrous oxide / acetylene flame at the 213.618 nm non-resonance line for phosphorus, and it is quite comparable with the LOD reported for GF AAS and ICP OES, except for those that are obtained in the vacuum-UV at 177.495 nm [82].

8.1.7 Determination of Sulfur in Cast Iron

Huang et al. [70] investigated in detail the determination of sulfur at CS molecular absorption lines in a fuel-rich air / acetylene flame, as described in Section 6.41. As sulfur may be present in a variety of chemical forms in a sample, the authors investigated sulfates, sulfites and sulfides under different conditions regarding their analytical sensitivity.

All standard solutions containing sulfate, such as H_2SO_4, $(NH_4)_2SO_4$ and $KHSO_4$ showed the same sensitivity, which was also essentially unchanged at different acid concentrations. Without acidification the standard solutions of Na_2SO_3 and $Na_2S_2O_3$ showed the same sensitivity as the sulfate solutions; however, in the presence of 1 % or higher acid concentration the sensitivity for sulfur was enhanced by up to a factor of two. The sensitivity for a standard solution of $(NH_4)_2S$, either acidified or not, was always higher by a factor of about two, however, upon addition of 1 % H_2O_2, the sensitivity was reduced to the same level as that of the sulfates, most likely because the sulfide was oxidized to the sulfate. This means, although the sulfide exhibits significantly higher sensitivity, any sulfur species present in the sample should be oxidized to sulfate during sample preparation in order to avoid any interference.

The authors applied the method for the determination of sulfur in two cast iron CRM (EURONOM, BAM, Berlin, Germany). They found a weak iron line at 258.045 nm in the vicinity of the 258.06 nm CS molecular line, which has been found to give the best performance (refer to Section 6.41). Although this iron line was very weak and did not cause any spectral overlap on the 258.06 nm CS line, as shown in Figure 8.12, it complicated background correction on the short wavelength side of this line. Therefore the CS line at 257.96 nm was used for this application. Because the final solution for measurement contained about 2 % m/v iron, the aspiration rate of the nebulizer was strongly influenced, so the authors used the analyte addition technique for calibration. The results obtained for the determination of sulfur in two cast iron CRM using the proposed method are shown in Table 8.2.

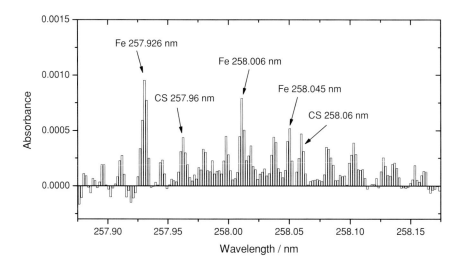

Figure 8.12: Absorbance of high iron concentrations in the vicinity of the CS band head at 258.06 nm

Table 8.2: Determination of sulfur in cast iron CRM (BAM, Berlin, Germany) using the CS molecular absorption at 257.96 nm and HR-CS F AAS with a fuel-rich air / acetylene flame

CRM	found value / % S	certified value / % S
EURONOM ZRM 428-2	0.105 ± 0.003	0.111 ± 0.002
EURONOM ZRM 484-1	0.229 ± 0.011	0.230 ± 0.004

8.2 Graphite Furnace Measurements

8.2.1 Method Development for Graphite Furnace Analysis

Obviously, the basic procedure and the measures taken in method development for GF AAS are identical for LS AAS and HR-CS AAS, as the same atomizers and control units are used. However, the special features of the latter technique, particularly the fact that the spectral environment of the analytical line becomes 'visible' at high resolution, facilitate method development tremendously. In addition, as dynamic signals are generated in the graphite atomizer, and absorbance is a function of time, the array detector truly introduces the wavelength resolution as a third dimension. Also, because of the dynamic character of all signals generated in the graphite furnace, the simultaneous double-beam system and the simultaneous background correction become of greatest importance, as even the fastest changes in absorbance over time cannot cause any signal distortion, as frequently observed in LS AAS.

The first step of a method development in GF AAS is usually an optimization of the pyrolysis and atomization temperatures by establishing pyrolysis and atomization curves using a matrix-free calibration solution as well as at least one representative sample or reference material. The pyrolysis curve exhibits the integrated absorbance signal obtained at a fixed atomization temperature as a function of the pyrolysis temperature, as shown schematically in Figure 8.13.

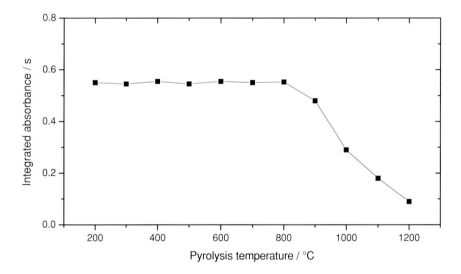

Figure 8.13: Typical pyrolysis curve for GF AAS; the integrated absorbance obtained at the optimum atomization temperature is plotted against the pyrolysis temperature

Usually the pyrolysis curve is characterized by a plateau up to the temperature where the analyte becomes volatile, and is lost in part during the pyrolysis stage. The atomization curve exhibits the integrated absorbance signal as a function of the atomization temperature using a fixed pyrolysis temperature. For volatile elements the atomization curve usually exhibits a maximum, followed by a decrease in integrated absorbance due to increasing diffusion losses at higher temperatures, as shown schematically in Figure 8.14. For less volatile analytes the atomization curve usually only reaches a plateau.

8. Specific Applications

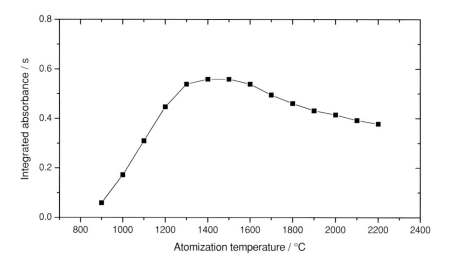

Figure 8.14: Typical atomization curve for GF AAS; the integrated absorbance is plotted against the atomization temperature at which it has been measured

The application of stabilized temperature platform furnace (STPF) conditions [133], particularly the atomization of the analyte from a platform in a transversely heated graphite tube, and the use of integrated absorbance (peak area) for signal evaluation are of main importance in order to reduce matrix effects to a minimum and ensure a good day-to-day reproducibility. Although spectral interferences are much less likely to occur in HR-CS AAS because of the high spectral resolution, and are much easier to correct for (see Section 5.2), it is advisable to use the highest possible loss-free pyrolysis temperature in order to remove as much of the matrix as possible prior to the atomization stage and to minimize potential non-spectral interferences. The absorbance pulse for volatile elements often extends over several seconds at the atomization temperature that corresponds to the maximum of the atomization curve so that often a 100–200 °C higher atomization temperature is preferred, with the advantage of a much 'sharper' signal. The atomization temperature should, however, not be chosen to be too high in order to avoid less volatile matrix constituents that are co-volatilized with the analyte and not sufficiently separated in time.

In LS AAS only the absorbance integrated over the emission profile of the radiation source is measured, so that in GF AAS an absorbance-over-time signal is obtained that consists of the analyte (atomic) absorption, and other absorption that might occur within that spectral range. The difficulty of distinguishing between the different types of absorption is one of the major limitations of LS AAS, which often complicates method development for graphite furnace application. A few practical examples are given in the following in order

8.2 Graphite Furnace Measurements

to demonstrate how the additional information available in HR-CS AAS can be used for a much more efficient method development and optimization of analytical parameters.

Thallium is one of the most difficult elements for GF AAS, at least when it has to be determined in complex matrices, as will be shown in more detail in Sections 8.2.6 and 8.2.7. Thallium forms a chloride that is volatile at temperatures above 400 °C and easily lost without being atomized, as the gaseous molecule TlCl is thermally very stable. Palladium, the most frequently used modifier to avoid such losses, has an absorption line close to the Tl line, causing a spectral interference. Iron, a frequent concomitant in many samples, also has an absorption line close to the Tl line. Finally, sulfur-containing samples produce a molecular absorption spectrum with a pronounced rotational fine structure, which is another source of spectral interferences [141].

Figure 8.15 (a), which shows the absorbance over time in the vicinity of the Tl resonance line at 276.787 nm during the atomization of a sediment sample at 2400 °C, illustrates perfectly the above-described problems [151]. The most prominent absorbance peak under these conditions at 276.752 nm is due to the matrix element iron. Obviously, in HR-CS AAS this iron line would not cause any interference, as it is far enough from the thallium line, but this figure clearly shows that the temperature program is not optimized

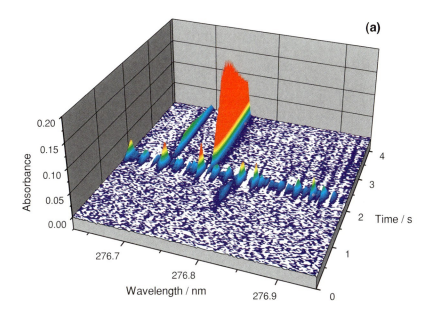

Figure 8.15: Time-resolved absorbance spectrum obtained for 0.1 mg PACS-2 Marine Sediment reference material around 276.787 nm; (a) atomization temperature 2400 °C;

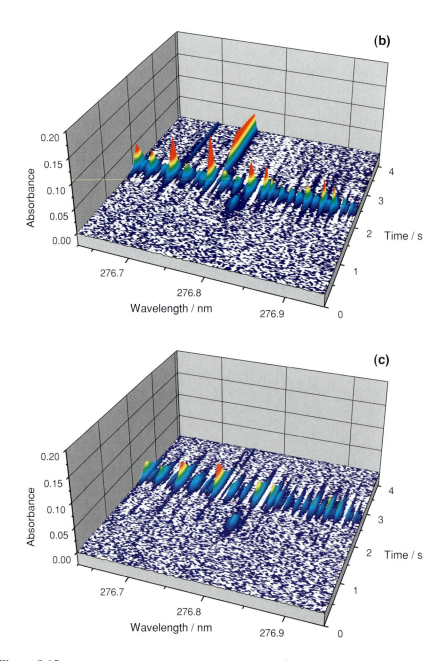

Figure 8.15: (continued); (b) atomization temperature 2000 °C; and (c) atomization temperature 1700 °C

for the determination of thallium. A reduction of the atomization temperature to 2000 °C (Figure 8.15 (b)) and 1700 °C (Figure 8.15 (c)), respectively, reduced and finally eliminated the iron absorption line. This is in contrast to the molecular absorption spectrum that follows the thallium absorption closely in time, and which could not be eliminated. The origin of this electron excitation spectrum is not yet fully identified, but it is certainly due to a sulfur-containing diatomic molecule; according to recent investigations by Katskov et al. [77, 87], the most likely source responsible for this absorption is the S_2 molecule.

One possibility that HR-CS AAS offers in order to cope with this potential interference is least-squares BC (see Section 5.2.2), in cases where the source of the molecular absorption is known and a reference spectrum can be produced. In the present example the reference spectrum could be generated by the vaporization of $KHSO_4$, which is shown in Figure 8.16.

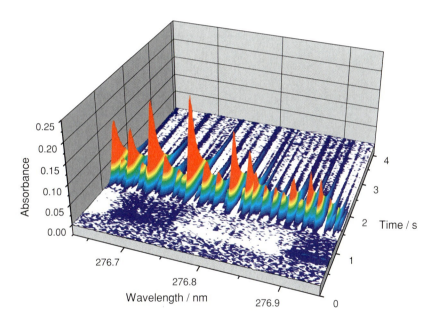

Figure 8.16: Molecular absorbance spectrum recorded for 0.01 mg $KHSO_4$ in the vicinity of 276.787 nm

The fact that only the thallium absorption line was left over after subtracting the reference spectrum, as can be seen in Figure 8.17, demonstrates that no other atomic or molecular absorption was recorded under these conditions within the spectral interval of interest.

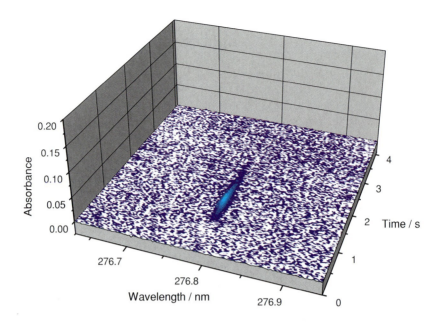

Figure 8.17: Residual absorbance signal for 0.1 mg PACS-2 Marine Sediment reference material around 276.787 nm after subtraction of the reference spectrum in Figure 8.16 from the spectrum shown in Figure 8.15 (c) using least-squares BC

A further optimization of the graphite furnace program is possible through time resolution, as vaporization of the different atomic and molecular species is often not simultaneous. This optimization is, in principle, also possible in LS GF AAS, however only in much coarser steps of usually only 0.1 s, and with the additional difficulty that it is notoriously difficult to distinguish between atomic and molecular absorption with this technique. In HR-CS AAS this distinction is much easier, as will be shown, and the time resolution for selecting the window for integration is about one order of magnitude better, depending only on the number of spectra recorded per second.

This optimization is shown for the same element thallium in a different matrix, coal, using direct solid sample analysis [130]. The overall absorbance pattern for this type of sample is shown in Figure 8.18. It is not significantly different from that obtained for the sediment samples, as coal also contains large amounts of iron and sulfur in the form of pyrite (FeS). The major difference is the very pronounced continuous background absorption at the beginning of the atomization stage. When a pyrolysis temperature of 600 °C is used, the coal matrix is not destroyed, and a very dense smoke and soot is generated at the beginning of the atomization cycle, causing an absorbance of about four. This means that

8.2 Graphite Furnace Measurements

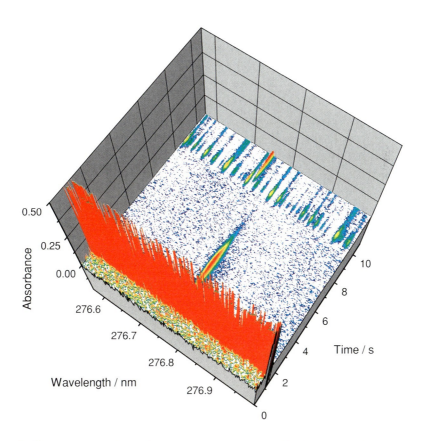

Figure 8.18: Time-resolved absorbance spectrum obtained for BCR 180 gas coal reference material in the vicinity of the thallium absorption line at 276.787 nm without a modifier; pyrolysis 600 °C, atomization 1700 °C

there is essentially no more radiation arriving at the detector, resulting in the extremely noisy signal in the initial phase of the atomization stage. However, even this extreme situation can be handled satisfactorily by the superior background correction capabilities of HR-CS AAS as illustrated in Figure 8.19 (a). For clarity, only the absorbance over time at the center pixel is shown for this example. Although an extreme baseline noise is generated by the excessive background absorption at the beginning of the atomization stage, the correction system works perfectly, as soon as the background absorbance drops to values around $A = 2$, making possible an interference-free evaluation of the thallium signal, which is perfectly separated in time from the background absorption. In addition, the pyrolysis temperature can be optimized further, as shown in Figure 8.19 (b), although the

8. Specific Applications

requirements for an optimization are less strict than in LS AAS. Increasing the pyrolysis temperature to 700 °C removes all residual problems, as the background absorbance reaches only values around $A = 2$, which are perfectly corrected. Figure 8.19 (b) also demonstrates that the simultaneous double-beam system eliminates any source noise, resulting in an extremely stable signal with an excellent SNR.

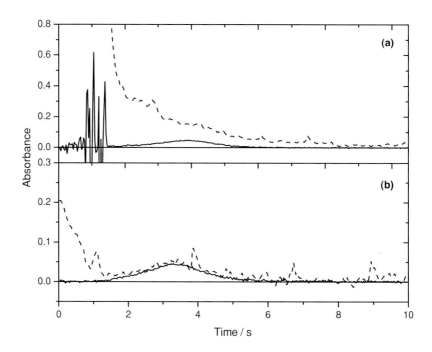

Figure 8.19: Absorbance over time recorded at the center pixel for the thallium line at 276.787 nm, for BCR 180 Gas Coal reference material, using direct solid sampling without modifier, without (dashed line) and with least-squares BC (solid line); atomization temperature 1700 °C; (a) pyrolysis temperature 600 °C; (b) pyrolysis temperature 700 °C

The atomization temperature can be optimized in a very similar way with the goal being to separate the atomic from the molecular absorption in time, which is shown in the following figures using the absorbance over time at the center pixel ± 2 pixels on each side. Figure 8.20 shows the optimization of the atomization temperature for thallium in coal using palladium as chemical modifier. In the case of HR-CS AAS this modifier can be used without problems, as the palladium absorption line is well separated from the thallium line. At an atomization temperature of 1600 °C (Figure 8.20 (a)) there is only the atomic absorbance signal for thallium visible, but the absorbance pulse extends over more than

5 s, which is far from optimum considering the SNR. At an atomization temperature of 1700 °C (Figure 8.20 (b)) the thallium signal becomes much sharper, but at the same time the structures of the sulfur molecule appear, overlapping at least in part with the atomic absorbance signal. At an atomization temperature of 1800 °C (Figure 8.20 (c)), finally, the overlap of the atomic and the molecular absorption is so strong that no more separation in time is possible, i.e. only least-squares BC could solve the problem.

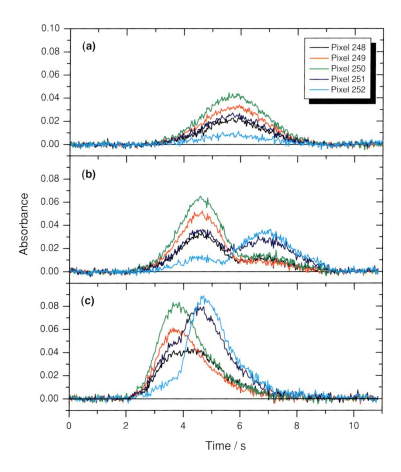

Figure 8.20: Absorbance over time recorded at the center pixel (250) for the thallium line at 276.787 nm and the two neighbor pixels at each side, for SARM 19 coal reference material, using direct solid sampling, and with palladium added in solution as modifier; pyrolysis temperature 700 °C; (a) atomization temperature 1600 °C; (b) atomization temperature 1700 °C; (c) atomization temperature 1800 °C

8. *Specific Applications*

In the absence of a modifier (Figure 8.21), however, the situation is significantly different, although at a temperature of 1600 °C a very similar, broad absorbance signal was obtained (which is not shown here), calling for a higher atomization temperature. This time, the molecular absorption at an atomization temperature of 1700 °C (Figure 8.21 (a)) was significantly lower and well separated from the thallium absorption, even at an atomization temperature of 1800 °C (Figure 8.21 (b)), although the molecular absorption was very pronounced, it was still separated in time from the atomic absorption signal. This might be considered as an example, where the use of a modifier actually has an undesired side effect, as it delays the atomization of the analyte, and hence causes the partial overlap of the atomic with the molecular absorption. Obviously this example should not be interpreted as a general recommendation not to use chemical modifiers, as each case has to be considered individually.

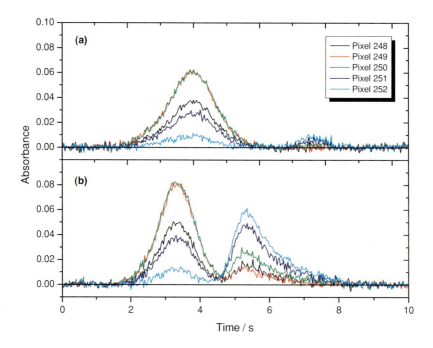

Figure 8.21: Same as Figure 8.20, but without the addition of a modifier; (a) atomization temperature 1700 °C; (b) atomization temperature 1800 °C

The time-resolved absorbance spectrum obtained for BCR 180 Gas Coal reference material in the vicinity of the thallium absorption line at 276.787 nm under optimized conditions is shown in Figure 8.22.

8.2 Graphite Furnace Measurements

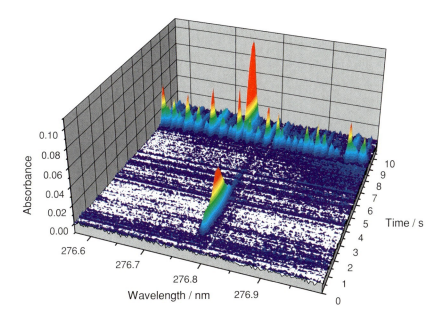

Figure 8.22: Time-resolved absorbance spectrum obtained for BCR 180 Gas Coal reference material in the vicinity of the thallium absorption line at 276.787 nm without a modifier; pyrolysis 700 °C, atomization 1700 °C

8.2.2 Direct solid sample analysis

The direct analysis of solid samples, independent of the analytical technique that is used for that purpose, has a number of distinct advantages over procedures that involve an acid digestion or other dissolution methods. Although the relative importance of the various aspects depends very much on the type of sample that has to be analyzed, the analyte that has to be determined and its concentration, and the specific requirements regarding speed of analysis, accuracy and precision, the following features are generally accepted:

1. Direct analysis of solid samples is faster as only a minimum of sample preparation is required; typical sample preparation steps, such as grinding and homogenization of heterogeneous materials, have to be carried out prior to a sample digestion.

2. Direct analysis of solid samples is more sensitive, as the analyte content is not diluted during sample preparation.

3. The risks of analyte loss and/or contamination are reduced to a minimum as essentially no reagents and only a minimum of laboratory ware is used.

8. Specific Applications

4. The use of corrosive and potentially hazardous reagents is avoided, reducing the waste problem tremendously, and contributing to the concept of 'green chemistry'.

Obviously there are also some disadvantages associated with the direct analysis of solid samples, such as difficulties in finding proper standards for calibration, and a greater imprecision due to sample inhomogeneity. The latter problem, however, is often counterbalanced by a better trueness of the results because of the reduced risk of analyte loss or contamination.

Graphite furnace AAS is particularly suited for the direct analysis of solid samples because of its discontinuous mode of operation, its capability to accommodate milligram masses, and even relatively large pieces of sample, and the long residence time of the vaporized sample in the absorption volume, which allows efficient atomization even of refractory compounds. Graphite atomizers have been used for the direct analysis of solid samples from the very beginning by L'vov [96] and Massmann [101]; an overview of the history and recent applications can be found in the book of Kurfürst [83]. Modern instrumentation greatly facilitates the introduction of solid samples into the graphite furnace, offering manual and automatic tools for sample handling, as shown in Figure 8.23. A prerequisite for a successful application of GF AAS for the direct analysis of solid samples is a consequent use of the STPF concept, and particularly of integrated absorbance for quantification in order to eliminate any kinetic effect on atom release, and hence on peak shape.

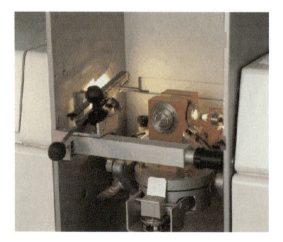 

Figure 8.23: Manual (left) and automatic (right) solid sampling accessories for GF AAS (Analytik Jena AG)

HR-CS AAS further enhances the applicability of GF AAS for the analysis of solid samples, predominantly because of its far superior background correction capabilities, including the least-squares BC. The greatly simplified method development, particularly in the presence of complex matrices, which has been discussed in Section 8.2.1, is a further contribution to facilitate direct solid sample analysis. The most important feature of HR-CS AAS for this application might however be that the emission intensity of the radiation source, and therefore also the source-related noise, are, to a first approximation, independent of the wavelength. As it is rather difficult to dilute solid samples, one of the elegant ways to cope with high analyte concentrations is the use of alternate, less sensitive analytical lines for their determination. In LS AAS, however, the intensity of these alternate lines, emitted by a hollow cathode lamp, was often relatively weak, resulting in a poor SNR and in a further deterioration of the precision. As this problem does not exist in HR-CS AAS, many more lines are available for analytical purposes without compromises. For a number of elements with a large number of absorption lines, even two or three lines of different sensitivity may be found within the selected spectral window, making possible the simultaneous use of more than one calibration curve, and hence a significant extension of the analytical working range.

8.2.3 Urine Analysis

Urine was one of the first real samples investigated by Heitmann and co-workers [58–60] using an HR-CS AAS instrument with a magnet at the furnace in order to study the effect of the magnetic field on the fine structure of the electron excitation spectra of gaseous molecules, such as PO, that originate from the urine matrix. The molecular absorbance spectra for PO without and with a magnetic field at the atomizer, and the difference between the two spectra are shown in Figures 8.24 and 8.25 in the vicinity of the wavelength for cadmium at 228.802 nm, and for palladium at 247.642 nm, respectively. In both cases the molecular spectrum of PO was affected by the magnetic field, indicating that BC errors might be expected when Zeeman-effect BC is used for this kind of determination. Similar problems were later reported for the determination of arsenic and selenium in urine [7]. It should only be mentioned for completeness that none of the other BC systems that are available for LS AAS, i.e., deuterium BC and high-current pulsing (Smith-Hieftje BC) are capable of handling such fine-structured background.

The wavelength- and time-resolved absorbance spectrum recorded for a human urine sample in the vicinity of the arsenic line at 193.696 nm is shown in Figure 8.26 (a), exhibiting a coincidence in time of the atomic and molecular absorption. This is a particularly difficult case, as there might be some molecular absorption underneath the analyte absorption line.

8. Specific Applications

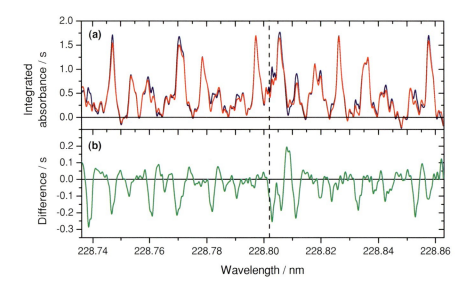

Figure 8.24: Absorbance spectrum of PO in the vicinity of the cadmium line at 228.802 nm (dashed line); (a) without (blue) and with magnetic field (red), and (b) difference of the spectra

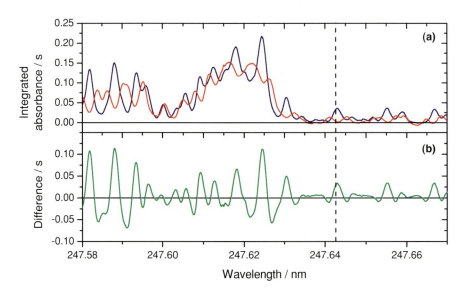

Figure 8.25: Absorbance spectrum of PO in the vicinity of the palladium line at 247.642 nm (dashed); (a) without (blue) and with magnetic field (red); (b) difference of the spectra

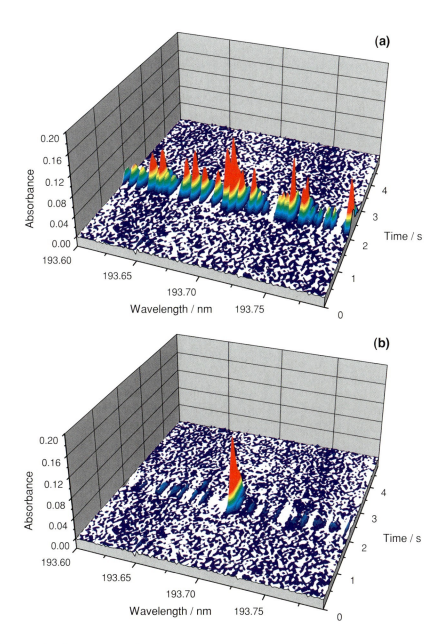

Figure 8.26: Time-resolved absorbance spectrum of a human urine sample in the vicinity of the arsenic absorption line at 193.696 nm; (a) without correction; (b) after subtraction of the spectra in Figure 8.27 using least-squares BC

8. Specific Applications

The time-integrated absorbance spectra of NaCl and PO (generated by atomization of $NH_4H_2PO_4$) within the same wavelength interval are shown in Figure 8.27, indicating that there is no direct overlap between atomic and molecular absorption in this case. This means that the spectral resolution of the instrument would have been sufficient to ensure an interference-free determination of arsenic in urine. Subtracting both spectra, using least-squares BC, resulted in the spectrum shown in Figure 8.26 (b), indicating complete removal of all matrix-related absorption within the considered spectral interval. The authors reported good agreement of their values with the certified values of an analyzed control urine [7].

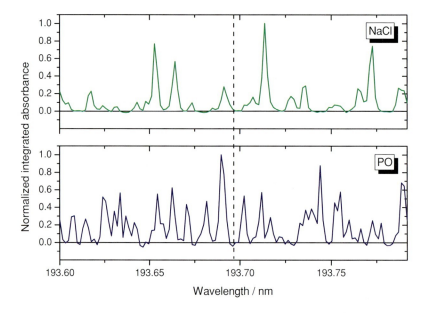

Figure 8.27: Time-integrated reference spectra of sodium chloride (50 μg NaCl) and PO (20 μg $NH_4H_2PO_4$) in the vicinity of the arsenic absorption line at 193.696 nm (dashed line)

In the case of cadmium in urine, only the molecular absorbance spectrum of PO caused interference, as shown in Figures 8.28 and 8.29. However, the situation in this case was particularly complex, firstly, because there was again a direct coincidence in time between the atomic and the molecular absorption (Figure 8.28 (a)), and secondly because the molecular absorbance spectrum of PO was particularly rich in fine-structure lines (Figure 8.29) that were at least in part overlapping with the atomic absorption line at 228.802 nm. This means that the only solution in this case is least-squares BC which, however, removed all molecular absorption perfectly, as shown in Figure 8.28 (b) [60].

8.2 Graphite Furnace Measurements

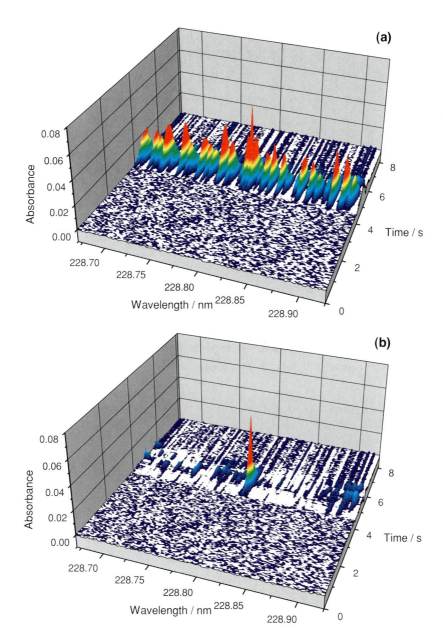

Figure 8.28: Time-resolved absorbance spectrum of a human urine sample in the vicinity of the cadmium absorption line at 228.802 nm; (a) without correction; (b) after subtraction of the spectra in Figure 8.29 using least-squares BC

8. Specific Applications

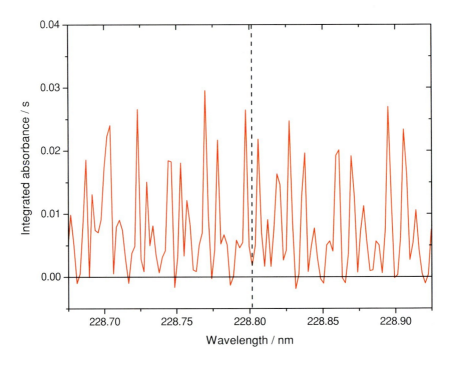

Figure 8.29: Time-integrated reference spectrum of PO (generated by the atomization of 20 μg NH$_4$H$_2$PO$_4$) in the vicinity of the cadmium absorption line at 228.802 nm (dashed line)

The determination of selenium in urine, at least at first glance, appeared to be somewhat less problematic, as the atomic absorption was separated in time from the molecular structures, as is depicted in Figure 8.30 (a). However, the molecular structures did not all appear at the same point in time, indicating that more than one gaseous molecule might be involved.

A detailed investigation revealed that, in addition to PO, NO, originating from urea (CH$_4$N$_2$O), was also contributing to molecular absorption, as shown in Figure 8.31. Subtraction of both spectra using least-squares BC resulted in a 'clean' spectrum containing the selenium atomic absorption only (Figure 8.30 (b)).

The temporal behavior of the corresponding correction factors provided by the least-squares BC algorithm, which are a gauge for the intensities of the reference spectra during sample atomization, is depicted in Figure 8.32. Again, a good agreement of the analytical results with certified values was obtained [7].

8.2 Graphite Furnace Measurements

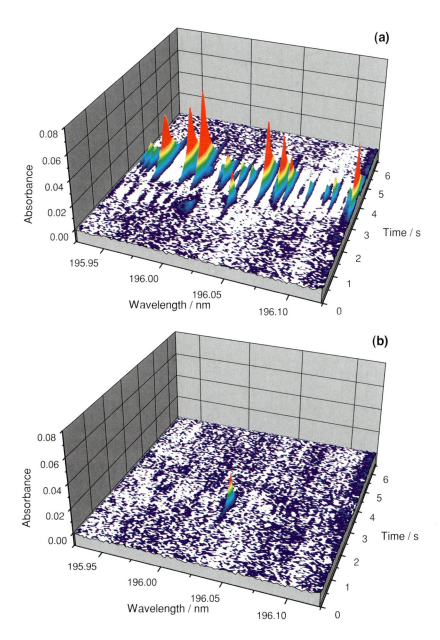

Figure 8.30: Time-resolved absorbance spectrum of a human urine sample in the vicinity of the selenium absorption line at 196.026 nm; (a) without correction; (b) after subtraction of the spectra in Figure 8.31 using least-squares BC

8. Specific Applications

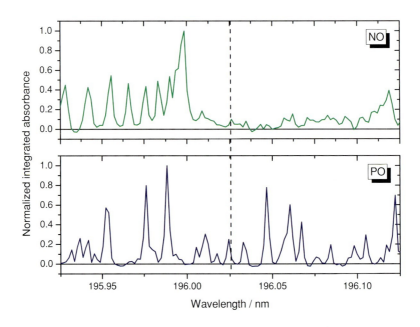

Figure 8.31: Time-integrated reference spectra of NO (100 µg urea, CH$_4$N$_2$O) and PO (20 µg NH$_4$H$_2$PO$_4$) in the vicinity of the selenium absorption line at 196.026 nm (dashed)

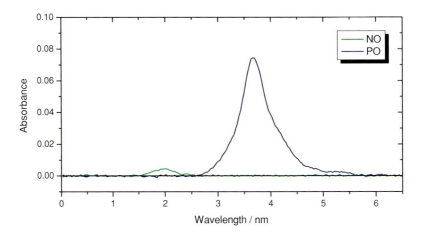

Figure 8.32: Temporal behavior of the correction factors given by the least-squares BC algorithm for the two reference spectra in Figure 8.31

8.2.4 Analysis of Biological Materials

Borges et al. [12] investigated the determination of lead in biological materials at the 217.001 nm line using direct solid sample analysis and ruthenium as permanent modifier. Figure 8.33 shows, in a very convincing manner, how HR-CS AAS can solve even very complex situations, as they were encountered in the determination of lead in BCR 186 Pig Kidney reference material. The 3D plot of Figure 8.33 (a) is dominated by the strong continuous absorbance signal at the beginning of the atomization cycle and the very pronounced molecular structures that follow soon after in time. The absorbance pulse for lead that appears slightly before the molecular absorption is barely visible. The erratic signal that precedes the atomization pulse is actually not the continuous absorption, which is obviously corrected automatically, and hence not visible.

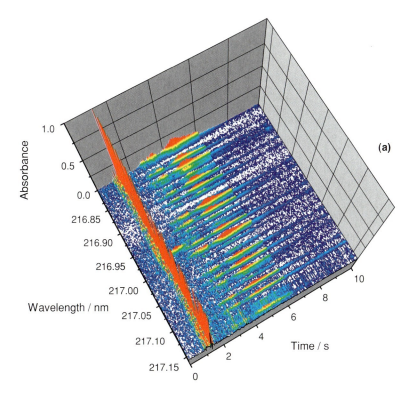

Figure 8.33: Determination of lead in BCR 186 Pig Kidney reference material using ruthenium permanent modifier; $T_{pyr} = 500\,°C$; $T_{at} = 1900\,°C$; (a) absorbance over time and wavelength in the vicinity of the lead resonance line at 217.001 nm;

8. Specific Applications

However, as a pyrolysis temperature of only 500 °C has been used in this case, the continuous background reached values of $A = 4$, which means that essentially no more radiation had arrived at the detector. What one can see in Figure 8.33 (a) is actually the excessive baseline noise that results from this situation.

The picture looks even more complex when the wavelength is integrated over time, obviously starting the integration only after the excessively noisy phase at the beginning of the atomization stage, as shown in Figure 8.33 (b). This plot is completely dominated by the molecular structures, and no absorbance signal at all is visible at the analytical wavelength of lead, which is indicated by the dotted line. However, the situation is actually much less complex than it looks. Firstly, the molecular structures are exclusively due to the PO molecule and an iron line at 216.885 nm, which, however, is too far from the analyte line to be of interest. The molecular absorption can obviously be easily corrected for by 'subtracting' the PO spectrum using least-squares BC.

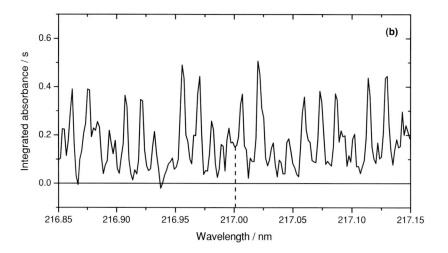

Figure 8.33: (continued); (b) time-integrated absorbance spectrum; lead wavelength at 217.001 nm is indicated by the dashed line; time delay after onset of the atomization cycle: 1.5 s;

But in the present case it was not even necessary to take any action because of the high spectral and time resolution of the equipment. When the absorbance over time, recorded at the center pixel of the lead absorption line is considered, which is shown in Figure 8.33 (c), it becomes obvious that actually none of the PO bands is overlapping with the analytical line. Hence it was only necessary to choose the beginning of the integration period in such a way that the noisy initial stage was excluded, and its end after the atomization pulse had returned to the baseline, as indicated in Figure 8.33 (c). In this way interference-free determination could be achieved, and accurate values for lead were obtained in a variety of biological materials without the need for any additional BC, using calibration against aqueous standards.

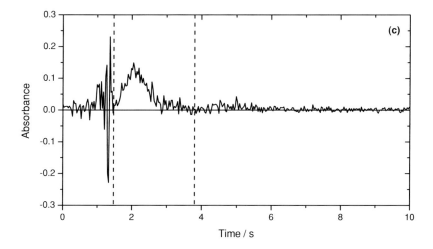

Figure 8.33: (continued); (c) absorbance over time recorded at the center pixel only; integration limits are indicated by dashed lines

The situation might actually be even more simplified for volatile elements that can be atomized satisfactorily at temperatures below 1700 °C, because the phosphates in biological materials, which are the source of the PO molecule and the spectral interference, are volatilized only above this temperature. Lepri et al. [89] determined cadmium in a number of biological reference materials, such as fish muscle, pig kidney, lichen, mussel tissue, and gull egg powder, using direct solid sample analysis. The samples were weighed in aliquots of 0.1 – 1 mg on a solid sampling graphite boat, which was introduced directly into the graphite tube. Iridium was used as a permanent modifier, allowing a maximum pyrolysis temperature of 800 °C. At pyrolysis temperatures below 700 °C, very high continuum background absorption preceded the analytical signal, which could not be completely separated in time. Structured background absorption due to the PO molecular absorption appeared at atomization temperatures higher than 1700 °C. An optimum pyrolysis temperature of 800 °C, and an atomization temperature of 1600 °C were therefore adopted for the determinations, eliminating the need for any additional BC, as shown in Figure 8.34. Calibration against aqueous solutions was carried out, resulting in good agreement with the certified values for cadmium, according to a t-test for a confidence level of 95 %. The relative standard deviations were lower than 7 %, demonstrating a good precision for solid sampling. The LOD was given as 1.3 pg Cd.

8. Specific Applications

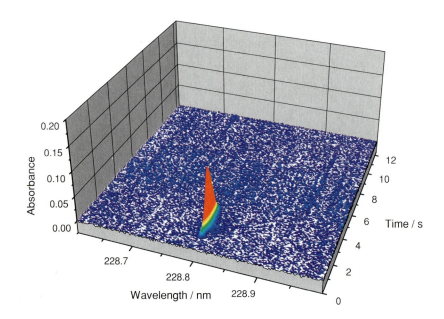

Figure 8.34: Time-resolved absorbance spectrum obtained for DORM-1 Dogfish Muscle reference material in the vicinity of the cadmium resonance line at 228.802 nm; pyrolysis temperature 800 °C; atomization temperature 1600 °C; iridium as permanent modifier; direct solid sample analysis

Ribeiro et al. [119] determined cobalt in a variety of biological samples, such as dogfish muscle, dogfish liver, lobster hepatopancreas, oyster tissue, bovine liver and human hair, using solubilization with tetramethylammonium hydroxide (TMAH) and direct solid sampling, and comparing LS AAS and HR-CS AAS. Alkaline treatment was carried out by adding about 250 mg of the sample to polyethylene flasks and 2 mL of 25 % m/v TMAH, with deionized water added up to the final volume of 15 mL. No additional sample treatment was applied for direct solid sample analysis. A pyrolysis temperature of 1000 °C, and an atomization temperature of 2300 °C were found to be optimum, and no chemical modifier was added for the determination of cobalt. The atomic absorbance signal of cobalt was always preceded by a molecular absorbance spectrum due to the PO molecule, as shown in Figure 8.35, and the molecular absorption was for all samples much lower in the case of the TMAH solutions compared to direct solid sample analysis. However, in the latter case, the atomic absorbance signal was better resolved in time from the molecular structures than in the case of the TMAH solutions. The signals in Figures 8.35 are quite noisy, which is due to the very low concentration of cobalt in this sample, and the resulting low signal of only about $A = 0.025$.

8.2 *Graphite Furnace Measurements*

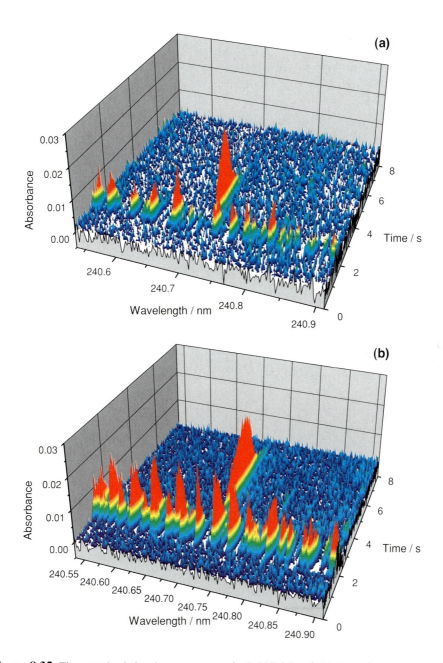

Figure 8.35: Time-resolved absorbance spectrum for DOLT-2 Dogfish Liver reference material in the vicinity of the cobalt absorption line at 240.725 nm; (a) solubilized in TMAH; (b) direct solid sample analysis

8. Specific Applications

The signals obtained for the same sample under the same conditions using LS AAS are shown in Figure 8.36. Obviously, the background signal measured with the deuterium lamp, which is integrated over a spectral interval of 0.2 nm, cannot be compared with that measured by HR-CS AAS, particularly as in the latter case any continuous background has already been eliminated.

The comparison is nevertheless interesting, because with LS AAS the background absorption appears to have the same magnitude for the TMAH solution and for solid sample analysis, although it is obvious from the measurements with HR-CS AAS that the fine-structured component is much greater in the latter case. This might also explain the much less pronounced over-correction in the case of the TMAH solution, although molecular and atomic absorption are not as well separated as in the case of solid sample analysis. This also explains why the results obtained with TMAH solutions using both procedures, LS and HR-CS AAS, were in good agreement with the certified values, according to the t-test for a 95 % confidence level, using calibration against aqueous standards.

In contrast to this, for direct analysis of solid samples, accurate results were obtained only with HR-CS AAS, whereas too high or too low values were obtained for several samples with LS AAS, obviously due to the much higher structured background absorption. The LOD of 5 ng/g obtained with HR-CS AAS was also about a factor of four better than that obtained with LS AAS.

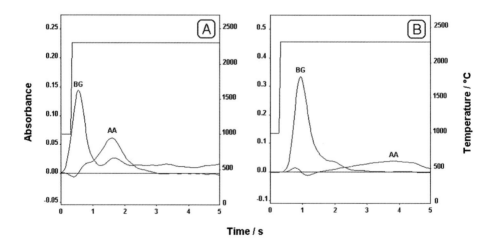

Figure 8.36: Absorbance-over-time recording for the atomic absorption (AA) and the background absorption (BG) for DOLT-2 Dogfish Liver reference material at the cobalt absorption line using conventional LS AAS; (a) solubilized in TMAH; (b) direct solid sample analysis

8.2.5 Analysis of Seawater

Salomon et al. [123] investigated in great detail the potential errors that could be encountered in the determination of nanomolar concentrations of aluminum in seawater, using conventional LS GF AAS with Zeeman-effect BG. In addition, HR-CS AAS was used as a diagnostic tool in order to identify the source of interference.

The authors decided to use the analytical line at 309.271 nm, because the seawater matrix caused a molecular absorption with a pronounced fine structure at the 396.153 nm line, as shown in Figure 8.37. This molecular absorption caused an interference when Zeeman-effect BC was used [123], whereas the analytical line at 309.271 nm was free from any molecular absorption. Experimental designs was used by the authors to optimize seven adjustable parameters of the spectrometer and furnace program, but they still found an influence of the lamp current and the slit width on the integrated absorbance of aluminum when seawater samples were analyzed, which was a clear indication for a spectral interference.

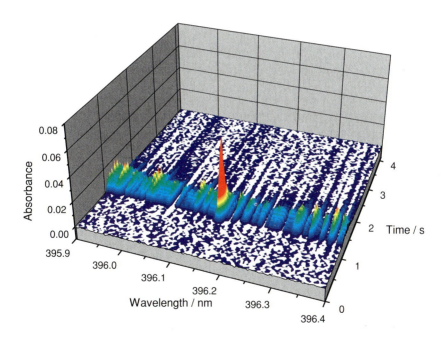

Figure 8.37: Time-resolved absorbance spectrum of a seawater sample in the vicinity of the aluminum absorption line at 396.153 nm; sample volume 25 μL; atomization temperature 2500 °C

8. Specific Applications

The authors pointed out correctly that, even when a seawater sample has gone through a pyrolysis stage of 20 s at 1720 °C, species such as Na_2S, Na_xO_y, CaO and MgO are still present in large amounts in the furnace. Then, the authors recorded both the emission spectrum of a single-element hollow cathode lamp and the absorbance spectrum of a seawater sample in the vicinity of the 309.271 nm aluminum line, as shown in Figure 8.38.

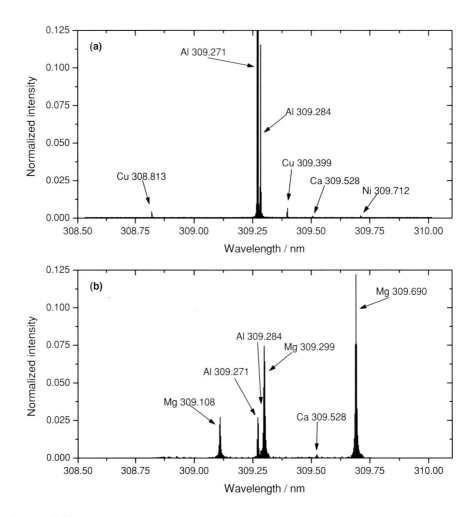

Figure 8.38: Time-integrated spectra in the vicinity of the aluminum line at 309.271 nm; (a) emission spectrum of an aluminum single-element hollow cathode lamp; lamp current: 23 mA; (b) absorbance spectrum of a seawater sample; sample volume 25 µL; atomization temperature 2500 °C

8.2 Graphite Furnace Measurements

They concluded that the presence of the calcium line at 309.528 nm in both spectra could obviously result in a small error. The authors also point out, considering the Zeeman-splitting in the range of 10–30 pm that the emission line of nickel partially coincides with the magnesium absorption line at 309.690 nm, and some overlap between the aluminum and magnesium absorption lines at 309.284 nm and 309.299 nm, respectively, occurred as well, resulting in the observed deviations. Although the authors did not use HR-CS AAS for the determination of aluminum in seawater, it is needless to say that the interferences detected for LS AAS with Zeeman-effect BC are absent in the former technique, so that an interference-free determination could be expected with an improved overall performance.

8.2.6 Analysis of Soils and Sediments

Welz et al. [151] investigated in detail the determination of thallium in marine sediment reference materials, a particularly complex topic, because of the volatility of TlCl, and the various spectral interferences that could be observed in the vicinity of the Tl line (refer to Section 8.2.1). The authors had previously investigated this using direct analysis of solid samples and Zeeman-effect BC [142], however they only succeeded in obtaining reliable results when ruthenium was used as permanent modifier, and a solution of ammonium nitrate was pipetted on top of the solid sample as an additional modifier. Method development turned out to be much easier in HR-CS AAS, and the spectral interference due to the sulfur content of the sediments could be completely removed using least-squares BC (refer to Section 8.2.1).

Using slurry sampling (100 mg of sediment in 5% v/v HNO_3) the authors found no statistical difference in the results for thallium without a modifier or with ruthenium and/or ammonium nitrate as modifiers, using aqueous standards for calibration, indicating complete removal of all interferences. The authors actually found some disadvantage in using the ruthenium modifier, as shown in Figure 8.39. Without the use of a modifier (Figure 8.39 (a)) the thallium atomic absorption peak was clearly separated in time from the fine-structured molecular absorption, and could be separated by just selecting the proper time interval for peak integration. In the presence of ruthenium, however, all absorption pulses, the atomic and the molecular structures, were broadened and, consequently, in part overlapping in time (Figure 8.39 (b)), making least-squares BC a necessity. The authors reported a detection limit of 0.02 μg/g Tl in sediment.

Silva et al. [131] investigated the determination of lead in soil and sediment slurries with ruthenium permanent modifier using LS AAS and HR-CS AAS, and comparing the analytical lines at 217.001 nm and 283.306 nm. It is well known that the former line provides about two-times higher sensitivity, but it is rarely used in conventional LS AAS because of its inferior SNR, its shorter linear working range, and its higher susceptibility

8. Specific Applications

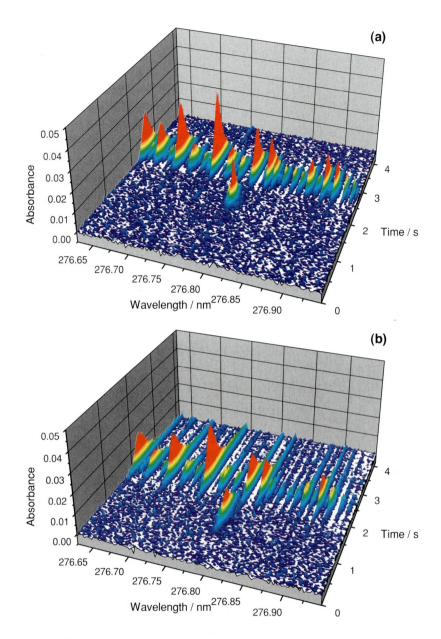

Figure 8.39: Time-resolved absorbance spectrum obtained for 0.1 mg MESS-1 Marine Sediment reference material in the vicinity of the thallium absorption line at 276.787 nm; (a) in 5 % v/v HNO_3; (b) in the presence of ruthenium as permanent modifier

to spectral interferences [150]. The authors confirmed the higher sensitivity of the line at 217.001 nm with characteristic mass values of 14 pg and 9 pg obtained with LS AAS and HR-CS AAS, respectively, compared to values of 24 pg and 18 pg, respectively at the 283.306 nm line. However, while LOD were similar around 2 µg/L for LS AAS and HR-CS AAS at the 283.306 nm line, this value improved to 1.0 µg/L for HR-CS AAS, and deteriorated to 2.5 µg/L for LS AAS at the 217.001 nm line. This result is according to expectations, as the radiation intensity at both lines is roughly the same in CS AAS, so that the better sensitivity also translates into a better LOD. The values determined for Pb in soil and sediment reference materials were in agreement within a confidence interval of 95 % for both instruments and both lines, although the RSD values were significantly higher for LS AAS using the 217.001 nm line.

Figure 8.40 shows for the example of BCR 142 Light Sandy Soil, that under optimized conditions, i.e. a pyrolysis temperature of 700 °C and an atomization temperature of 1700 °C, only the atomic absorption pulse, and no background absorption appeared within both wavelength intervals, except for the well-known iron line at 216.885 nm. The latter one, however, is too far from the lead line to cause any difficulties, and in addition it only appeared during the cleaning stage at the end of the furnace program.

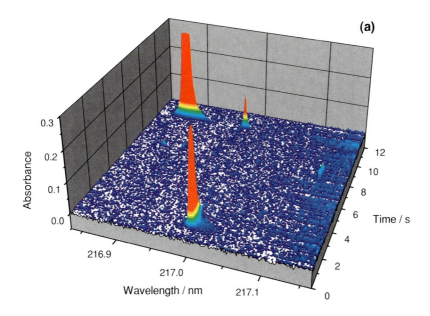

Figure 8.40: Time-resolved absorbance spectrum obtained for BCR 142 Light Sandy Soil reference material in the presence of ruthenium as permanent modifier; pyrolysis 700 °C, atomization 1700 °C; (a) in the vicinity of the lead absorption line at 217.001 nm;

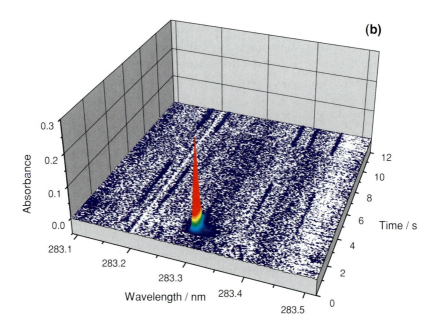

Figure 8.40: (continued); (b) in the vicinity of the lead absorption line at 283.306 nm

8.2.7 Analysis of Coal and Coal Fly Ash

Silva et al. [130] investigated in detail the determination of thallium in coal, as burning of coal is one of the major anthropogenic sources of this toxic element in our environment; an estimated 1100 tons of thallium are emitted into the atmosphere per year worldwide due to this process [141]. Direct analysis of solid samples was considered most appropriate due to the difficulties in bringing coal into solution, and the associated risk of losing thallium during this process due to the high volatility of its chloride. Another problem with this determination is that there is not a single coal reference material available with a certified value for thallium, making validation of methods difficult.

When solid coal is introduced into the graphite furnace, a minimum pyrolysis temperature of about 700 °C is necessary in order to remove the majority of the matrix prior to the atomization stage. A very dense smoke with peak absorbance values of 4–5 develops at the beginning of atomization when lower pyrolysis temperatures are used, resulting in an excessively noisy signal, because essentially no more radiation reaches the detector, as was shown in Figure 8.18 (refer to Section 8.2.1). However, when using HR-CS AAS, it was possible, even under these extreme conditions, to separate the analyte signal from the background absorption in time, as has been shown in Figure 8.19 (a) in Section 8.2.1, and to make an undisturbed measurement of the thallium absorption. Fortunately, however, thallium was apparently stabilized by the coal matrix itself, permitting a pyrolysis temperature of at least 700 °C without any analyte losses, and with significantly reduced background absorption, as shown in Figure 8.19 (b).

The optimization of the atomization temperature for this application has already been shown in Section 8.2.1 (Figures 8.20 and 8.21), and turned out to be quite critical, as atomic and molecular absorption overlap rapidly in time with increasing atomization temperature. When a chemical modifier, either palladium in solution or ruthenium as permanent modifier, was used the separation of the different signals in time became very difficult, so that either a very low atomization temperature had to be chosen, which resulted in significantly broadened atomization pulses, or least-squares BC had to be applied to correct for the molecular absorption. Without any modifier, however, atomic and molecular absorption remained well separated in time, at least up to an atomization temperature of 1700 °C, which is quite similar to the experience with the determination of thallium in marine sediments (refer to Section 8.2.6). It is quite remarkable that accurate results could be obtained for thallium in coal using direct solid sample analysis without the addition of any modifier and calibration against aqueous standards, compared with values that have been obtained by ETV-ICP-MS using isotopic dilution and analyte addition for calibration. The authors also analyzed a coal fly ash reference material, under the same conditions that were used for coal analysis, reaching the recommended value for thallium. The LOD for thallium in coal following this procedure was 0.01 mg/kg.

Borges et al. [13] investigated the determination of lead in coal following the same procedure, i.e. direct analysis of solid samples, without the use of a modifier, and calibration against aqueous standards. The most sensitive resonance line at 217.001 nm was selected for this determination, as with HR-CS AAS this line also provides a better SNR. Very similar to the determination of thallium, a pyrolysis temperature of 700 °C was sufficient to eliminate the continuous background due to the coal matrix prior to the atomization stage, and a molecular absorption with a pronounced fine structure, as well as a very strong absorption at the iron line at 216.885 nm, appeared when atomization temperatures higher than 1700 °C were used, as shown in Figure 8.41 (a). Although there was no molecular absorption band in the immediate neighborhood of the lead line and, in addition, the time limits for integration of the atomic absorbance pulse could be selected in such a way as to exclude any molecular absorption, it was found advantageous to use a lower atomization temperature, as all the molecular structures could be eliminated in this way, as shown in Figure 8.41 (b), and the results obtained for CRM were within the 95 % confidence interval of the certificate, using aqueous standards for calibration. The LOD for lead in coal was determined to be 0.1 mg/kg.

8. Specific Applications

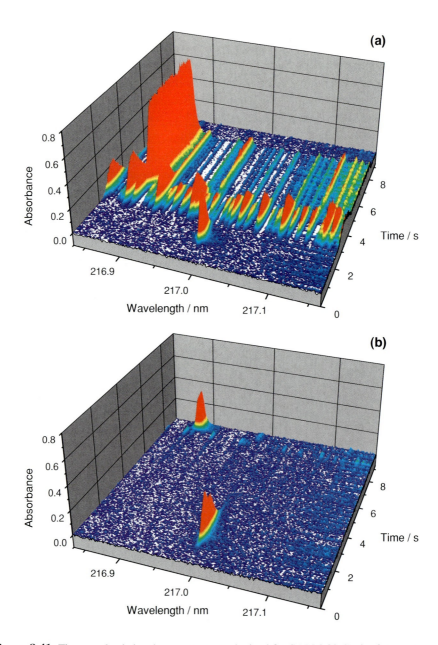

Figure 8.41: Time-resolved absorbance spectrum obtained for SARM-20 Coal reference material in the vicinity of the lead absorption line at 217.001 nm without a modifier; pyrolysis 700 °C; (a) atomization temperature 2000 °C; (b) atomization temperature 1700 °C

Silva et al. [132] also investigated the determination of cadmium in a variety of American, European and South African coal reference materials using direct analysis of solid samples and iridium as permanent modifier. Without a modifier cadmium was lost from aqueous solutions at pyrolysis temperatures above 300 °C, and from coal samples above 500–600 °C, depending on the specific coal. In the presence of iridium, cadmium in aqueous solution could be stabilized up to a pyrolysis temperature of 800 °C, and in coal samples to 600–900 °C, depending again on the specific coal. A small background signal preceded the atomic absorption signal of cadmium when a pyrolysis temperature of only 600 °C could be used, which did, however, not present any problem for the BC system, as can be seen in Figure 8.42. No molecular structures were observed during the atomization stage when an optimized atomization temperature of 1600 °C was used.

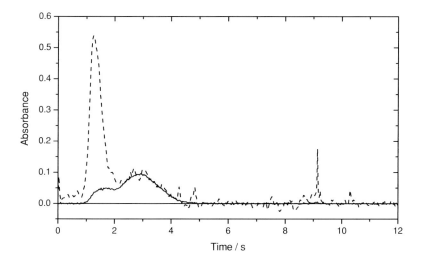

Figure 8.42: Absorbance over time without (dashed line) and with (solid line) automatic correction for continuous background for cadmium in SARM-19 Coal reference material at 228.802 nm; pyrolysis temperature: 600 °C, atomization temperature: 1600 °C; direct solid sample analysis with Ir as permanent modifier

However, higher atomization temperatures could be used without problem, as the analyte signal was in any case well resolved from the molecular absorption structures, as shown in Figure 8.43. The authors also investigated the influence of the sample mass introduced into the graphite tube on the precision, an important question in the direct analysis of solid samples. The authors found no significant influence on precision as long as the sample mass was higher than 0.4 mg, and the inhomogeneity of the investigated reference materials became a problem only for a sample mass below 0.3 mg.

8. Specific Applications

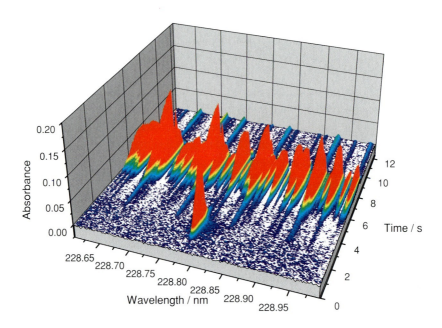

Figure 8.43: Time-resolved absorbance spectrum for cadmium in SARM-20 coal in the vicinity of 228.802 nm; pyrolysis temperature: 800 °C, atomization temperature: 1700 °C; direct solid sample analysis with Ir as permanent modifier

Bianchin et al. [11] described the determination of cadmium in coal using slurry sampling GF AAS and comparing LS AAS and HR-CS AAS. The authors investigated iridium, rhodium and ruthenium, alone and together with tungsten as permanent modifiers. These authors also arrived at the conclusion that iridium was the most appropriate modifier, however, a tungsten and iridium mixed modifier gave even better results, stabilizing cadmium up to 800 °C, and making possible calibration against aqueous standards.

8.2.8 Analysis of Crude Oil

Crude oil is a particularly complex and difficult matrix, but GF AAS has been used quite frequently, at least to determine elements such as nickel and vanadium, which are present in crude oil in relatively high concentration, making possible a reasonably high dilution of the crude oil matrix. Nickel and vanadium are particularly important elements, as their concentration and the ratio of the two elements are actually borehole specific, and make it possible to identify the origin of a crude oil. In addition, nickel and vanadium are both serious catalyst poisons, and may cause undesirable side reactions in refinery operations.

Vanadium in addition causes corrosion problems, e.g. in the combustion chamber of power plants, and inhalable dusts or aerosols of nickel that might be generated during combustion of oil are classified as hazardous substances owing to their carcinogenic and mutagenic effects.

Vale et al. [143], analyzing nickel in crude oil using LS GF AAS, observed an unusual slight increase in sensitivity in the pyrolysis curves for pyrolysis temperatures below 1000 °C for crude oil, whereas no such increase was found for the calibration solutions, as shown in Figure 8.44. Originally this increase in sensitivity was interpreted as an arti-

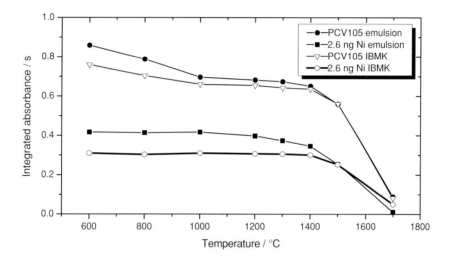

Figure 8.44: Pyrolysis curves for a 2.6 ng nickel standard solution (Ni salt of 2-ethylhexanoic acid) in base mineral oil, and a crude oil sample, diluted with IBMK and as oil-in-water emulsions, respectively, using LS GF AAS

fact due to some non-corrected background, as no determination was possible at pyrolysis temperatures lower than 600 °C due to excessive background absorption. However, when the authors used HR-CS AAS in order to identify the potential spectral interference, they discovered that there was only one additional absorption line within the spectral range of interest, which has been identified as another nickel line, but no molecular absorption that could have caused an increase in the analyte absorbance, as shown in Figure 8.45. This second nickel line actually proved to be very useful for HR-CS AAS as it made possible the simultaneous determination of nickel with both lines at significantly different sensitivity, as shown in Figure 8.46, extending the working range by one order of magnitude. This was particularly useful, as the concentration of nickel in crude oil may well change within three orders of magnitude, depending on the origin of the oil.

8. Specific Applications

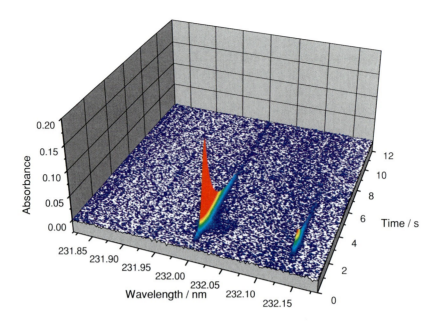

Figure 8.45: Time-resolved absorbance spectrum obtained for a Brazilian crude oil sample as oil-in-water emulsion in the vicinity of the nickel absorption line at 232.003 nm using a pyrolysis temperature of 600 °C, and an atomization temperature of 2400 °C

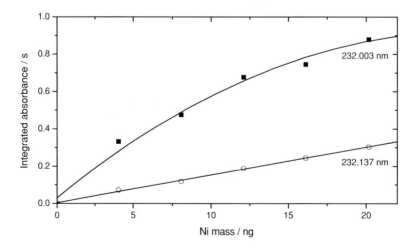

Figure 8.46: Calibration curves for nickel in oil, obtained simultaneously using the absorption lines at 232.003 nm and 232.137 nm, shown in Figure 8.45

In a next step the authors used the superior BC capability of HR-CS AAS in order to investigate the phenomenon at even lower pyrolysis temperatures. Although a very rapidly changing background signal appeared under these conditions early in the atomization stage, which exceeded values of $A = 2$, it could be corrected without any problems by the simultaneous BC system, as shown in Figure 8.47.

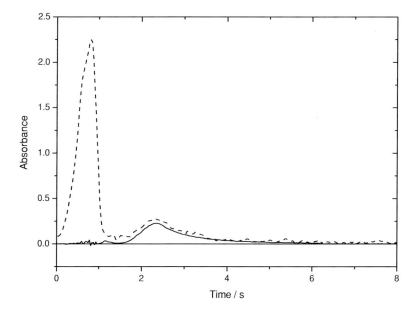

Figure 8.47: Absorbance over time recorded at the center pixel of the nickel absorption line at 232.003 nm for a Brazilian crude oil sample as oil-in-water emulsion, without (dashed line) and with (solid line) background correction, using a pyrolysis temperature of 400 °C, and an atomization temperature of 2400 °C

In addition, the atomic absorption signal of nickel was well separated in time from any molecular absorption, and it was even possible to use a pyrolysis temperature of 300 °C, which opened completely new diagnostic investigations. Under these conditions the authors discovered a very significant increase in sensitivity for nickel up to 400 °C, as shown in Figure 8.48.

Very similar behavior has also been discovered for vanadium in later work [144], as also shown in Figure 8.48. The only reasonable explanation for this behavior is that there are at least two significantly different types of nickel and vanadium compounds present in the crude oil, one that is volatile and lost at pyrolysis temperatures higher than 400 °C,

8. Specific Applications

and another type that is thermally stable up to pyrolysis temperatures of about 1300 °C for nickel and 1600 °C in the case of vanadium. Most likely the volatile compounds are low molecular weight, non-polar nickel and vanadyl porphyrins, whereas the stable compounds are polar, salt-like non-porphyrins.

The authors found that up to 50 % of the nickel and 40 % of the vanadium present in crude oil might be lost at temperatures above 400 °C, a loss that cannot be detected by conventional LS AAS. It might therefore be suspected that much data reported in the literature using LS GF AAS are wrong because of this phenomenon.

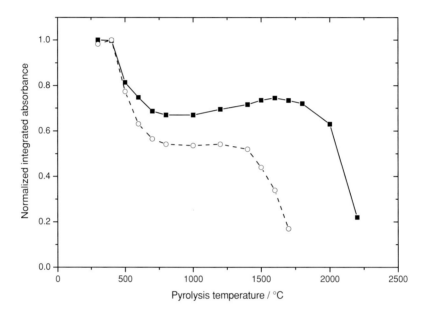

Figure 8.48: Pyrolysis curves for nickel (dashed line) and vanadium (solid line) in a Brazilian crude oil sample as oil-in-water emulsion, using HR-CS GF AAS, measured at 232.003 nm and 318.540 nm, respectively

These authors also investigated ways to avoid these low-temperature losses of nickel and vanadyl porphyrins, and found that 20 µg of reduced palladium, deposited onto the platform before sample introduction, efficiently avoided all losses, and stabilized all nickel and vanadium compounds in oil up to 1300 °C and 1600 °C, respectively [144], as shown in Figure 8.49.

8.2 Graphite Furnace Measurements

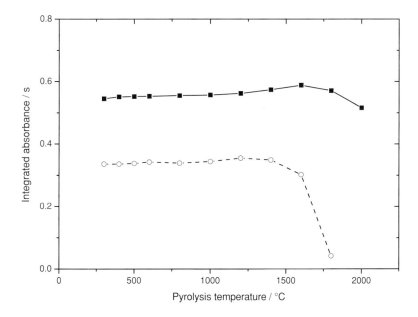

Figure 8.49: Pyrolysis curves for nickel (dashed line) and vanadium (solid line) in a Brazilian crude oil sample as oil-in-water emulsion, using 20 µg of reduced palladium as modifier, using HR-CS GF AAS, measured at 232.003 nm and 318.540 nm, respectively

8.2.9 Determination of Arsenic in Aluminum

Riley [120, 121] reported, already in 1982, a potential overlap of two aluminum lines at 193.471 nm and 193.583 nm with the most sensitive line for arsenic at 193.696 nm; this was confirmed by Slavin and Carnrick [135] a few years later. Martinsen et al. [100] reported a strong interference from aluminum on arsenic when they used an instrument with deuterium BC, which almost disappeared when they changed to Zeeman-effect BC. This was confirmed by Flajnik-Rivera and Delles [34], who found that as little as 10 mg/L Al gave a positive signal on the 193.696 nm arsenic wavelength, corresponding to 4 µg/L As when deuterium BC was used, whereas a ten times higher concentration of 100 mg/L Al was necessary to produce the same effect with Zeeman-effect BC. Schlemmer et al. [124] repeated the experiment later using two different atomic absorption spectrometers, both with Zeeman-effect BC, and found a positive bias caused by 100 mg/L (2 µg) Al corresponding to 3 µg/L As and 1.5 µg/L As, respectively, and they could remove the interference by the addition of 7.5 µg Mg as chemical modifier.

8. Specific Applications

Schlemmer et al. [124] continued this research using HR-CS AAS, where the nature of the background absorption becomes apparent, as shown in Figure 8.50. It is quite obvious from the structure of the 'peak' that the interference is not due to an atomic line, but to some molecular absorption, although the kind of molecule that causes the absorption is yet unknown. It also becomes clear that part of the problems observed in LS AAS is the rapid change of the absorbance with time, which is difficult to follow with sequential correction systems.

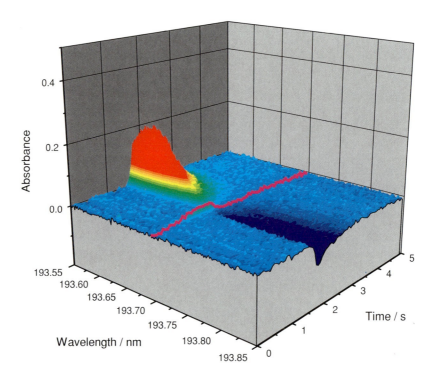

Figure 8.50: Background absorbance recorded over wavelength and time caused by 2 μg Al in the vicinity of the arsenic line at 193.696 nm (solid line); atomization temperature 2300 °C

In HR-CS AAS it is only necessary to select two reference pixels, the so-called background correction pixels (BCP), as described in Section 5.2.1, symmetrically at both sides of the arsenic wavelength in order to correct for this spectral interference, as shown in Figure 8.51.

8.2 Graphite Furnace Measurements

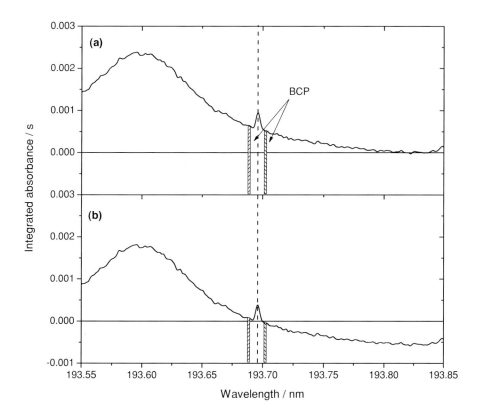

Figure 8.51: Selection of two background correction pixels (BCP) in order to correct for the background absorption of 2 μg Al in the vicinity of the arsenic line at 193.696 nm (dashed line); atomization temperature 2300 °C; (a) without correction; (b) with BCP correction

Nevertheless, the more elegant way to correct for the 'sloping' background becomes obvious from Figure 8.52. When selecting a certain number of sampling points over the recorded spectral interval, the molecular absorption can be fitted to a polynomial (refer to Figure 8.52 (a)). Subsequently, the application of a least-squares fitting algorithm, which is described in detail in Section 5.2.1, will bring the absorbance signal to the zero line, as shown in Figure 8.52 (b). In this way, the molecular absorption phenomenon is completely eliminated without any residual artifact in near vicinity of the absorption line, making possible the determination of very low arsenic concentrations even in the presence of a 100 000-fold excess of aluminum.

8. Specific Applications

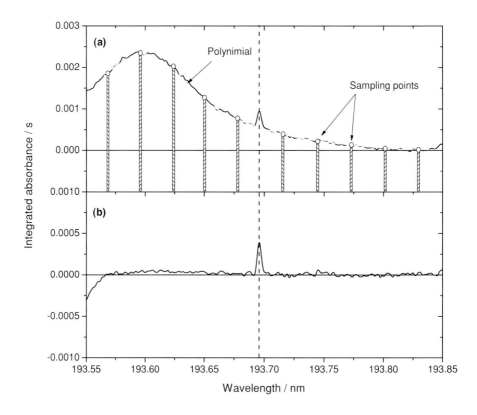

Figure 8.52: Integrated absorbance for 20 pg As in the presence of 2 μg Al (same as in Figure 8.51) when using a polynomial fit (red line) for least-squares BC; atomization temperature 2300 °C; (a) selection of sampling points; (b) after correction for sloping background

9. Outlook

The most obvious area for future development relates to simultaneous HR-CS AAS for routine analytical work. Unfortunately, the use of state-of-the-art CCD array detectors for HR-CS AAS, placed in the focal plane of echelle spectrographs with a two-dimensional spectrum pattern, is restricted at present to research applications. The problem arises from the well-known trade-off between the required pixel number and the resulting read-out time of common CCD arrays. The higher the pixel number of the array detector, which is strongly correlated with the simultaneously recorded wavelength interval, the longer it takes to read out the full image with the required dynamic range. Furthermore, the ratio between the spectrum illumination time, resulting from the high radiance of the continuum source, and the detector read-out time also gets worse with increasing detector area. Even though there is ongoing progress in the field of solid-state technology, an appropriate array detector for HR-CS AAS is not available today. In order to quantify this problem, the essential number of detector pixels in the case of simultaneous recording of broad wavelength ranges should be estimated roughly by the following approximations.

In the first step one should estimate the total amount of bandwidth increments which have to be recorded. It has been pointed out in Section 2.1.5 that for HR-CS AAS the width of the instrumental profile $\Delta\lambda_{instr}$ should depend on the wavelength λ in a form which is well met by echelle spectrometers with:

$$\Delta\lambda_{instr} = \frac{\lambda}{R}, \qquad (9.1)$$

where R is the instrumental resolving power. The total amount of simultaneously recorded wavelength increments M can be ascertained by integrating the number m of increments per nm via the valid wavelength range. Following Equation 9.1 it will be assumed that $m = 1/\lambda_{instr} = R/\lambda$. This results in:

$$M = \int_{\lambda_{min}}^{\lambda_{max}} m \, d\lambda = \int_{\lambda_{min}}^{\lambda_{max}} \frac{R}{\lambda} \, d\lambda . \qquad (9.2)$$

9. Outlook

The integration delivers in the known manner:

$$M = R \ln \frac{\lambda_{max}}{\lambda_{min}} . \tag{9.3}$$

Taking into account, that the entrance slit-determined width of the instrumental profile $\Delta\lambda_{instr}$ should be sampled by at least two pixels, the calculation delivers the number of required detector pixels to be $M_{pix} = 2\, R \ln(\lambda_{max}/\lambda_{min})$. Unfortunately this is far from reality, because the arrangement of the spectral orders in a two-dimensional echelle spectrum corresponding to Figure 3.12 is not strictly regular in relationship to the chessboard geometry of a solid-state detector. Becker-Ross and Florek [6] have given an estimation for the detector width, necessary for recording an echelle spectrum with prism cross dispersion. Following this approach, and assuming a slit height of two times the slit width ($h_{en} = 2s_{en}$), one gets a useful approximation for the essential number of pixels:

$$M_{pix} \approx 10\, \frac{\lambda_{max}}{\lambda_{min}}\, R \ln \frac{\lambda_{max}}{\lambda_{min}} . \tag{9.4}$$

For the considered wavelength range from 190 nm to 850 nm and for the preferable instrumental resolving power $R = 75\,000$ Equation 9.4 delivers $M_{pix} \approx 5$ Mega pixel.

This value appears to be quite common, when for example the market for digital cameras is considered. Since about 2004, the leading detector companies, such as Fairchild Imaging Inc. (Milpitas, USA), e2v Technologies (Chelmsford, UK), and DALSA Inc. (Waterloo, Canada), offer sensors with up to 85 Mega pixels and with a typical pixel size between 8 μm x 8 μm and 15 μm x 15 μm. The read-out frequency can be enhanced up to 200 MHz with a multi-directional read-out technology. A typical DALSA detector with 6.3 Mega pixels delivers 20 frames per second, a magnitude that is veritably interesting for AAS measurements. Without doubt, these innovative products are promising beginnings.

However, one has to take into account that the pixel size of these sensors is usually small, resulting in low saturation capacity, which is correlated to an inadequately reduced dynamic range for CS AAS, as has been shown in Section 3.3.

Furthermore, the balancing of the detector illumination is a crucial topic in simultaneous HR-CS AAS. To reach the full dynamic range, which is essential for low LOD over the total simultaneously recorded spectral range, the CCD image detector has to be saturated almost uniformly. This is not practical for the typical radiance distribution of available continuum sources, not even for the hot-spot xenon lamp described in this book.

The way out of this dilemma seems to be the random access to pixel sequences or actually to single detector pixels. In the future, active pixel arrays with one read-out amplifier at every pixel, which are still under development for the mass market, could overcome the limitations and might be the basis for the final step to simultaneous HR-CS AAS. Alternatively, specially designed segmented detectors, which combine the advantages of CCD detection with fast and flexible read-out, could be used successfully. But segmented detectors, as they have been applied by Harnly and co-workers [52], are not available in a sufficient variety, without even considering their actual prohibitive price.

In 1996 Harnly [50] was still optimistic when he concluded that 'multi-element AAS will ultimately be successful because it will offer figures of merit which are comparable or superior to those available today for single element AAS', and 'a CS AAS instrument ...would provide analytical capabilities comparable to ICP-MS with considerably less complexity'. In 1999, however, he became more pessimistic [54] when he provokingly concluded: 'The cost of development and construction of the ideal detector would be expensive. In the current economic climate, where development of technology for future products is sacrificed for quarterly profits, the cost of the detector may be sufficient to block the development of multi-element CS AAS, regardless of the enhanced analytical capabilities'.

The authors of this book are more optimistic and share Harnly's opinion of 1996, expecting that the rapid development in the wide field of solid-state detectors will provide an affordable solution in the foreseeable future. Even if there are still a lot of technical problems that have to be solved, and the costs are not really calculable at this point in time, it can be ascertained that instrumentation for simultaneous HR-CS AAS is no longer pure speculation.

Independent of the sequential or simultaneous detection capability, a very interesting aspect for the application of HR-CS AAS using existing instrumentation is certainly the possibility to determine elements via their diatomic molecules in flames as well as in graphite furnaces. The examples for P via PO and S via CS, which have been presented in Sections 8.1.6 and 8.1.7 may be only the beginning of various similar investigations. Especially non-metals, such as the halogens, will surely be at the center of interest.

Another area, which has not even been mentioned in this book, is the technique of chemical vapor generation with atomization in heated quartz tubes. Although this technique is generally considered to be relatively free of spectral interferences [150], there is relatively little known about the mechanism of mutual interferences of hydride-forming elements, to give only one example. HR-CS AAS may well be a very useful tool in elucidating these mechanisms through the identification of gaseous reaction products formed in the atomizer.

Last but not least, in spite of the triumphal procession of ICP-MS in modern analytical chemistry, an additional reference method which is based preferably on a different physical principle of signal generation is imperatively desirable to confirm or to disprove precarious results. From this point of view alone, HR-CS AAS, when its intrinsic potential is fully exploited, will, in the opinion of the authors, play an important role in future laboratory equipment.

References

[1] N. Allard, J. Kielkopf, Rev. Mod. Phys. 54 (1982) 1103.

[2] M. Baker, M. Denton, in: J. Sweedler, K. Ratzlaff, M. Denton, *Charge-Transfer Devices in Spectroscopy*, VCH, Weinheim, 1994.

[3] T.W. Barnard, M.J. Crockett, J.C. Ivaldi, P.L. Lundberg, D.A. Yates, P.A. Levine, D.J. Sauer, Anal. Chem. 65 (1993) 1231.

[4] H. Becker-Ross, S. Florek, R. Tischendorf, G.R. Schmecher, J. Anal. At. Spectrom. 10 (1995) 61.

[5] H. Becker-Ross, S. Florek, U. Heitmann, R. Weisse, Fresenius J. Anal. Chem. 355 (1996) 300.

[6] H. Becker-Ross, S. Florek, Spectrochim. Acta Part B 52 (1997) 1367.

[7] H. Becker-Ross, S. Florek, U. Heitmann, J. Anal. At. Spectrom. 15 (2000) 137.

[8] H. Becker-Ross, S. Florek, H. Franken, B. Radziuk, M. Zeiher, J. Anal. At. Spectrom. 15 (2000) 851.

[9] H. Becker-Ross, M. Okruss, S. Florek, U. Heitmann, M.D. Huang, Spectrochim. Acta Part B 57 (2002) 1493.

[10] H. Becker-Ross, S. Florek, U. Heitmann, M.D. Huang, M. Okruss, FACSS XXX, Fort Lauderdale, USA, Book of Abstracts (2003) 208.

[11] L. Bianchin, F.L. Sanchez, M.B. Dessuy, M.G.R. Vale, W.N.L. dos Santos, S.L.C. Ferreira, B. Welz, 8th Rio Symposium on Atomic Spectrometry, Paraty, Brazil, Book of Abstracts (2004) 60.

[12] D.L.G. Borges, *Direct determination of Lead in Biological Samples by High-Resolution Continuum Source Atomic Absorption Spectrometry*, Master thesis, Universidade Federal de Santa Catarina, Florianópolis, Brazil, 2005.

[13] D.L.G. Borges, 8th Rio Symposium on Atomic Spectrometry, Paraty, Brazil, Book of Abstracts (2004) 28.

[14] P.W.J.M. Boumans, J.J.A.M. Vrakking, Spectrochim. Acta Part B 39 (1984) 1239.

[15] G.R. Carnrick, W.B. Barnett, W. Slavin, Spectrochim. Acta Part B 41 (1986) 991.

[16] J.L. Carper, At. Absorpt. Newslett. 9 (1970) 48.

[17] P.R.M. Correira, P.V. Oliveira, J.A. Gomes-Neto, J.A. Nóbrega, J. Anal. At. Spectrom. 19 (2004) 917.

[18] H.M. Crosswhite, G.H. Dieke, C.S. Legagneur, J. Opt. Soc. Am. 45 (1955) 270.

[19] A. Corney, *Atomic and Laser Spectroscopy*, Oxford University Press, Oxford, 1977.

[20] H.M. Crosswhite, *The Hydrogen Molecule Wavelength Tables of H.G. Dieke*, Wiley-Interscience, 1972.

[21] R.F. Crow, J.D. Connolly, Prog. Anal. Spectrosc. 1 (1978) 347.

[22] A.J. Curtius, G. Schlemmer, B. Welz, J. Anal. At. Spectrom. 2 (1987) 115.

[23] M. Czerny, A.F. Turner, Z. f. Physik 61 (1930) 792.

[24] G. Daminelli, D.A. Katskov, R.M. Mofolo, P. Tittarelli, Spectrochim. Acta Part B 54 (1999) 669.

[25] G. Daminelli, D.A. Katskov, R.M. Mofolo, T. Kantor, Spectrochim. Acta Part B 54 (1999) 683.

[26] L. de Galan, W.W. McGee, J.D. Winefordner, Anal. Chim. Acta 37 (1967) 436.

[27] G.R. Dulude, J.J. Sotera, Can. Res. 15 (1982) 21.

[28] B. Edlen, Metrologia 2 (1966) 71.

[29] R.C. Elsner, J.D. Winefordner, Anal. Chem. 44 (1972) 698.

[30] V.A. Fassel, V.G. Mossotti, Anal. Chem. 35 (1963) 252.

[31] V.A. Fassel, V.G. Mossotti, W.E.L. Grossman, R.N. Kniseley, Spectrochim. Acta 22 (1966) 347.

[32] R. Fernando, B.T. Jones, Spectrochim. Acta Part B 49 (1994) 615.

[33] K.G. Fernandes, M. Moraes, J.A. Gomes-Neto, J.A. Nóbrega, P.V. Oliveira, Analyst 127 (2002) 157.

[34] C. Flajnik-Rivera, F. Delles, Am. Environ. Lab. 8/3 (1996) 24.

[35] S. Florek, H. Becker-Ross, J. Anal. At. Spectrom. 10 (1995) 145.

[36] S. Florek, H. Becker-Ross, T. Florek, Fresenius J. Anal. Chem. 355 (1996) 269.

[37] S. Florek, U. Heitmann, M. Schütz, H. Becker-Ross, XXX Colloquium Spoectroscopicum Internationale, Melbourne, Australia, Book of Abstracts (1997) 284.

[38] C.W. Frank, W.G. Schrenk, C.E. Meloan, Anal. Chem. 39 (1967) 534.

[39] C. Frigge, E. Jackwerth, Spectrochim. Acta Part B 47 (1992) 787.

[40] R.C. Fry, M.B. Denton, Appl. Spectrosc. 33 (1979), 393.

[41] J.H. Gibson, W.E. Grossman, W.D. Cooke, Appl. Spectrosc. 16 (1962), 47.

[42] A.Kh. Gilmutdinov, J.M. Harnly, Spectrochim. Acta Part B 53 (1998) 1003.

[43] V.L. Ginsburg, G.I. Satarina, Zavodsk. Lab. 31 (1964) 249.

[44] J.M. Harnly, T.C. O'Haver, B. Golden, W.R. Wolf, Anal. Chem. 51 (1979) 2007.

[45] J.M. Harnly, J.S. Kane, N.J. Miller-Ihli, Appl. Spectrosc. 36 (1982) 637.

[46] J.M. Harnly, N.J. Miller-Ihli, T.C. O'Haver, Spectrochim. Acta Part B 39 (1984) 305.

[47] J.M. Harnly, Anal. Chem. 58 (1986), 933A.

[48] J.M. Harnly, J. Anal. At. Spectrom. 8 (1993) 317.

[49] J.M. Harnly, Spectrochim. Acta Part B 48 (1993) 909.

[50] J.M. Harnly, Fresenius J. Anal. Chem. 355 (1996) 501.

[51] J.M. Harnly, C.M.M. Smith, B. Radziuk, Spectrochim. Acta Part B 51 (1996) 1055.

[52] J.M. Harnly, C.M.M. Smith, D.N. Wichems, J.C. Ivaldi, P.L. Lundberg, B. Radziuk, J. Anal. At. Spectrom. 12 (1997) 617.

[53] J.M. Harnly, R.E. Fields, Appl. Spectrosc. 51 (1997) 334A.

[54] J.M. Harnly, J. Anal. At. Spectrom. 14 (1999) 137.

[55] J.M. Harnly, A. Gilmutdinov, M. Schuetz, J. Murphy, J. Anal. At. Spectrom. 16 (2001) 1241.

[56] G.H. Harrisson, J. Opt. Soc. Am. 39 (1949) 522.

[57] M. Hedrich, U. Rösick, P. Brätter, R.L. Bergmann, K.E. Bergmann, in: B. Welz, *5. Colloquium Atomspektrometrische Spurenanalytik*, Bodenseewerk Perkin-Elmer, Überlingen (1989) 879.

[58] U. Heitmann, M. Schütz, H. Becker-Ross, S. Florek, Spectrochim. Acta Part B 51 (1996) 1095.

[59] U. Heitmann, M. Schütz, H. Becker-Ross, S. Florek, FACSS XXIII, Kansas City, USA, Book of Abstracts (1996) 173.

[60] U. Heitmann, M. Schütz, H. Becker-Ross, S. Florek, XXX Colloquium Spoectroscopicum Internationale, Melbourne, Australia, Book of Abstracts (1997) 47.

[61] U. Heitmann, H. Becker-Ross, GIT Labor-Fachzeitschrift 7 (2001) 728.

[62] U. Heitmann, S. Florek, H. Becker-Ross, 7th Rio Symposium on Atomic Spectrometry, Florianópolis, Brazil, Book of Abstracts (2002) 200.

[63] G. Herzberg, *Molecular Spectra and Molecular Structure: I. Spectra of Diatomic Molecules*, Van Nostrand Reinhold Company, New York, 1950.

[64] G.M. Hieftje, J. Anal. At. Spectrom. 4 (1989) 117.

[65] J.A. Holcombe, J.M. Harnly, Anal. Chem. 58 (1986) 2606.

[66] G.C. Holst, *CCD Arrays, Cameras and Displays*, SPIE Optical Engineering Press, Bellingham, 1996.

[67] M.D. Huang, H. Becker-Ross, S. Florek, U. Heitmann, CANAS t'01 – Colloquium Analytische Atomspektroskopie, Freiberg, Germany, Book of Abstracts (2001) 70.

[68] M.D. Huang, H. Becker-Ross, S. Florek, U. Heitmann, M. Okruss, *Application of High-Resolution PO Molecular Absorption Spectra to the Direct Determination of Phosphorus using a Continuum Source Atomic Absorption Spectrometer and an Air-Acetylene Flame*, J. Anal. At. Spectrom. (2005) submitted.

[69] M.D. Huang, H. Becker-Ross, S. Florek, U. Heitmann, M. Okruss, *Investigation and Elimination of the Suppression Effect of Calcium and Magnesium on the PO Molecular Absorption Signal in the Direct Determination of Phosphorus using a Continuum Source Atomic Absorption Spectrometer and an Air-Acetylene Flame*, J. Anal. At. Spectrom. (2005) submitted.

[70] M.D. Huang, H. Becker-Ross, S. Florek, U. Heitmann, M. Okruss, *Application of High-Resolution CS Molecular Absorption Spectra to the Direct Determination of Sulphur using a Continuum Source Atomic Absorption Spectrometer and an Air-Acetylene Flame*, J. Anal. At. Spectrom. (2005) submitted.

[71] J.D. Ingle, S.R. Crouch, *Spectrochemical Analysis*, Prentice Hall Inc., New Jersey, 1988.

[72] N.P. Ivanov, N.A. Kozyreva, Zh. Analit. Khim. 19 (1964) 1266.

[73] J.F. James, R.S. Sternberg, *The Design of Optical Spectrometers*, Chapman and Hall, London, 1969.

[74] B.T. Jones, B.W. Smith, J.D. Winefordner, Anal. Chem. 61 (1989) 1670.

[75] D.A. Katskov, R.M. Mofolo, P. Tittarelli, Spectrochim. Acta Part B 55 (2000) 1577.

[76] D.A. Katskov, R.M. Mofolo, P. Tittarelli, Spectrochim. Acta Part B 56 (2001) 57.

[77] D.A. Katskov, M. Lemme, P. Tittarelli, Spectrochim. Acta Part B 59 (2004) 101.

[78] P.N. Keliher, C.C. Wohlers, Anal. Chem. 46 (1974) 682.

[79] G. Kirchhoff, Phil. Mag. 20 (1860) 1.

[80] G. Kirchhoff, R. Bunsen, Phil. Mag. 20 (1860) 89.

[81] G. Kirchhoff, R. Bunsen, Phil. Mag. 22 (1861) 329.

[82] K. Krengel-Rothensee, U. Richter, P. Heitland, J. Anal. At. Spectrom. 14 (1999) 699.

[83] U. Kurfürst, *Solid Sample Analysis*, Springer, Berlin, 1998.

[84] R. Kurucz, CD-ROM No. 23, Harvard-Smithsonian Center for Astrophysics, 1995.

[85] P. Larkins, J. Anal. At. Spectrom. 7 (1992) 265.

[86] A. Le Bihan, J.Y. Cabon, C. Elleouet, Analusis 20 (1992) 601.

[87] M. Lemme, D.A. Katskov, P. Tittarelli, Spectrochim. Acta Part B 59 (2004) 115.

[88] W. Lenz, Z. Phys. 80 (1933) 423.

[89] F.G. Lepri, A.F. Silva, B. Welz, A.J. Curtius, U. Heitmann, 8th Rio Symposium on Atomic Spectrometry, Paraty, Brazil, Book of Abstracts (2004) 152.

[90] S.A. Lewis, T.C. O'Haver, J.M. Harnly, Anal. Chem. 56 (1984), 1066.

[91] D.R. Lide, *CRC Handbook of Chemistry and Physics*, CRC Press, Boca Raton, 1994.

[92] J.N. Lockyer, *Studies in Spectrum Analysis*, Appleton, London, 1878.

[93] E. Lundberg, W. Frech, J.M. Harnly, J. Anal. At. Spectrom. 3 (1988) 1115.

[94] B.V. L'vov, A.D. Khartsyzov, Zh. Prikl. Spektrosk. 11 (1969) 9.

[95] B.V. L'vov, Spectrochim. Acta Part B 33 (1978) 153.

[96] B.V. L'vov, Spectrochim. Acta Part B 39 (1984) 149.

[97] B.V. L'vov, V.G. Nikolaev, E.A. Norman, L.K. Polzik, M. Mojika, Spectrochim. Acta Part B 41 (1986) 1043.

[98] B.V. L'vov, Spectrochim. Acta Part B 45 (1990) 633.

[99] B.V. L'vov, Spectrochim. Acta Part B 54 (1999) 1637.

[100] I. Martinsen, B. Radziuk, Y. Thomassen, J. Anal. At. Spectrom. 3 (1988) 1013.

[101] H. Massmann, Fresenius Z. Anal. Chem. 225 (1967) 203.

[102] H. Massmann, Z. El Gohary, S. Güçer, Spectrochim. Acta Part B 31 (1976) 399.

[103] W.W. McGee, J.D. Winefordner, Anal. Chim. Acta 37 (1967) 429.

[104] N.J. Miller-Ihli, T.C. O'Haver, J.M. Harnly, Anal. Chem. 56 (1984), 176.

[105] A. Mitchell, M. Zemansky, *Resonance Radiation and Excited Atoms*, Cambridge University Press, Cambridge, 1972.

[106] G.P. Moulton, T.C. O'Haver, J.M. Harnly, J. Anal. At. Spectrom. 4 (1989) 673.

[107] G.P. Moulton, T.C. O'Haver, J.M. Harnly, J. Anal. At. Spectrom. 5 (1990) 145.

[108] NIST National Institute of Standard and Technology (USA), *NIST Atomic Spectra Database*, http://physics.nist.gov/cgi-bin/AtData/main_asd, Gaithersburg, 2004.

[109] G.J. Nitis, V. Svoboda, J.D. Winefordner, Spectrochim. Acta Part B 27 (1972) 345.

[110] T.C. O'Haver, J.D. Messman, Prog. Anal. Spectrosc. 9 (1986) 483.

[111] K.E.A. Ohlsson, W. Frech, J. Anal. At. Spectrom. 4 (1989) 379.

[112] A.P. Oliveira, J.A. Gomes-Neto, J.A. Nóbrega, P.V. Oliveira, Talanta 64 (2004) 334.

[113] O. Oster, *Zum Selenstatus in der Bundesrepublik Deutschland*, Universitätsverlag Jena, Jena, 1992.

[114] F. Paschen, Ann. der Physik 50 (1916) 901.

[115] R.W.B. Pearse, A.G. Gaydon, *The Identification of Molecular Spectra*, Chapman & Hall, London, 1950 & 1976.

[116] D.V. Posener, Austr. J. Phys. 12 (1959) 184.

[117] W.H. Press, B.P. Flannery, S.A. Teukolsky, W.T. Vetterling, *Numerical Recipes in Pascal*, Cambridge University Press, Cambridge, 1989.

[118] B. Radziuk, N.P. Romanova, Y. Thomassen, Anal. Commun. 36 (1999) 13.

[119] A.S. Ribeiro, M.A. Vieira, A.F. Silva, D.L.G. Borges, B. Welz, A.J. Curtius, U. Heitmann, *Determination of cobalt in biological samples by line-source and high-resolution continuum source graphite furnace atomic absorption spectrometry using solid sampling or alkaline treatment*, Spectrochim. Acta Part B 60 (2005) accepted.

[120] K.W. Riley, At. Spectrosc. 3 (1982) 120.

[121] K.W. Riley, Analyst 109 (1984) 181.

[122] A.N. Saidel, W.K. Prokofjew, S.M. Raiski, *Spektraltabellen*, VEB Verlag Technik, Berlin, 1955.

[123] S. Salomon, P. Giamarchi, A. Le Bihan, H. Becker-Ross, U. Heitmann, Spectrochim. Acta Part B 55 (2000) 1337.

[124] G. Schlemmer, A. Glomb, U. Heitmann, 7th Rio Symposium on Atomic Spectrometry, Florianópolis, Brazil, Book of Abstracts (2002) 7.

[125] G. Schlemmer, H. Gleisner, T.Z. Guo, R. Herkle, U. Heitmann, 8th Rio Symposium on Atomic Spectrometry, Paraty, Brazil, Book of Abstracts (2004) 7.

[126] K.P. Schmidt, H. Becker-Ross, S. Florek, Spectrochim. Acta Part B 45 (1990) 1203.

[127] D. J. Schroeder, *Astronomical Optics*, Academic Press, San Diego, 1987.

[128] M. Schütz, *Untersuchungen über den Einfluß von Untergrundabsorptionen in der Kontinuumstrahler-Atomabsorptionsspektrometrie und ihre Korrektur*, PhD thesis, Technische Universität Berlin, Berlin, Germany, 1997.

[129] M. Schütz, J. Murphy, R.E. Fields, J.M. Harnly, Spectrochim. Acta Part B 55 (2000) 1895.

[130] A.F. Silva, D.L.G. Borges, B. Welz, M.G.R. Vale, M.M. Silva, A. Klassen, U. Heitmann, Spectrochim. Acta Part B 59 (2004) 841.

[131] J.S.A. Silva, D. Budziak, D.L.G. Borges, A.J. Curtius, B. Welz, U. Heitmann, 8th Rio Symposium on Atomic Spectrometry, Paraty, Brazil, Book of Abstracts (2004) 184.

[132] A.F. Silva, D.L.G. Borges, F.G. Lepri, B. Welz, A.J. Curtius, U. Heitmann, *Determination of cadmium in coal using solid sampling graphite furnace high-resolution continuum source atomic absorption spectrometry, calibration against aqueous standards and Ir as a permanent modifier*, J. Anal. Atom. Spectrom. (2005) submitted.

[133] W. Slavin, D.C. Manning, G.R. Carnrick, At. Spectrosc. 2 (1981) 137.

[134] W. Slavin, G.R. Carnrick, Spectrochim. Acta Part B 39 (1984) 271.

[135] W. Slavin, G.R. Carnrick, CRC Crit. Rev. Anal. Chem. 19 (1988) 95.

[136] C.M.M. Smith, J.M. Harnly, Spectrochim. Acta Part B 49 (1994) 387.

[137] C.M.M. Smith, J.M. Harnly, J. Anal. At. Spectrom. 10 (1995) 187.

[138] W. Snelleman, Spectrochim. Acta Part B 23 (1968) 403.

[139] A.R. Striganov, N.S. Sventitskii, *Tables of Spectral Lines of Neutral and Ionized Atoms*, IFI/Plenum, 1998.

[140] J. Sweedler, K. Ratzlaff, M. Denton, *Charge-Transfer Devices in Spectroscopy*, VCH, Weinheim, 1994.

[141] M.G.R. Vale, B. Welz, Spectrochim. Acta Part B 57 (2002) 1821.

[142] M.G.R. Vale, M.M. Silva, B. Welz, R. Nowka, J. Anal. Atom. Spectrom. 17 (2002) 38.

[143] M.G.R. Vale, I.C.F. Damin, A. Klassen, M.M. Silva, B. Welz, A.F. Silva, F.G. Lepri, D.L.G. Borges, U. Heitmann, Microchem. J. (2004) 131.

[144] M.G.R. Vale, B. Welz, M.M. Silva, F.G. Lepri, I.C.F. Damin, W.N.L. dos Santos, U. Heitmann, 8th Rio Symposium on Atomic Spectrometry, Paraty, Brazil, Book of Abstracts (2004) 21.

[145] C. Veillon, P. Merchant, Appl. Spectrosc. 27 (1973) 361.

[146] A. Walsh, Spectrochim. Acta 7 (1955) 108.

[147] A. Walsh, Anal. Chem. 46 (1974) 698A.

[148] R. Weiße, *Untersuchung der instrumentellen und spektroskopischen Voraussetzungen für die Atomabsorptionsspektroskopie unter Verwendung einer kontinuierlichen Strahlungsquelle*, PhD thesis, Technische Universität Berlin, Berlin, Germany, 1996.

[149] B. Welz, Z. Grobenski, Appl. At. Spectrom. 10 (1978) 1.

[150] B. Welz, M. Sperling, *Atomic Absorption Spectrometry*, 3rd edition, Wiley-VCH, Weinheim, 1999.

[151] B. Welz, M.G.R. Vale, M.M. Silva, H. Becker-Ross, M.D. Huang, S. Florek, U. Heitmann, Spectrochim. Acta Part B 57 (2002) 1043.

[152] R. Wennrich, W. Frech, E. Lundberg, Spectrochim. Acta Part B 44 (1989) 239.

[153] G. Wibetoe, F.J. Langmyhr, Anal. Chim. Acta 165 (1984) 87.

[154] G. Wibetoe, F.J. Langmyhr, Anal. Chim. Acta 176 (1985) 33.

[155] G. Wibetoe, F.J. Langmyhr, Anal. Chim. Acta 186 (1986) 155.

[156] G. Wibetoe, F.J. Langmyhr, Anal. Chim. Acta 198 (1987) 81.

[157] D.N. Wichems, R.E. Fields, J.M. Harnly, J. Anal. At. Spectrom. 13 (1998) 1277.

References

[158] W.L. Wiese, M.W. Smith, B.M. Glennon, *Atomic Transition Probabilities*, Vol. I & II, NBSDS, Washington, 1966 & 1969.

[159] E. Wunderlich, W. Hädeler, Fresenius Z. Anal. Chem. 281 (1976) 1068.

[160] X.F. Yin, G. Schlemmer, B. Welz, Anal. Chem. 59 (1987) 1462.

[161] A.T. Zander, T.C. O'Haver, P.N. Keliher, Anal. Chem. 48 (1976) 1166.

Acknowledgment

The authors of this book are particularly grateful for the financial support given by Analytik Jena AG, Germany, which made possible the design and construction of the prototype HR-CS AAS instruments at ISAS - Institute for Analytical Sciences, Department Berlin, Germany, that have been used for the detailed technical and analytical evaluation of this technique. Most of the work described in this book, particularly the investigations about the analytical feasibility of HR-CS AAS at the Department of Chemistry, Federal University of Santa Catarina (UFSC), Brazil, and at ISAS, Department Berlin, would not have been possible without that support.

The authors would also like to acknowledge the contribution of M. Okruss, who was responsible for realization and interpretation of spectrograph measurements, particularly for the molecular overview spectra and the corresponding theoretical background, of M.D. Huang and R. Schmecher, who were collecting the basic data for Chapter 6, as well as for the determination of phosphorus and sulfur using molecular absorption bands. G. Wesemann, L. Mollwo and R. Tischendorf were responsible for the mechanical and electronic design, respectively, and H. Bräutigam carried out the lamp measurements. Among the various users of this equipment, D.L.G. Borges, F.G. Lepri, A.F. Silva and M.G.R. Vale had the most important contribution to the analytical evaluation of HR-CS AAS at UFSC in Brazil, predominantly in the field of GF AAS.

One of the authors (B. Welz) is grateful to Coordenação e Aperfeiçoamento de Pessoal de Nível Superior (CAPES) for a research scholarship.

Index

Aberration 11
Absolute analysis 74
Absorbance 18
    integrated 22, **84**, 226, 236
    maximum 22
    minimum detectable 18
    negative 189, 211
    peak (maximum) 21
Absorbance noise 19, 68–72
Absorbance spectrum
    AgH molecule 155
    air/acetylene flame 187
    AlCl molecule 158
    AlF molecule 160
    AlH molecule 162
    AsO molecule 164
    CN molecule 166
    CS molecule 169
    CuH molecule 173
    GaCl molecule 176
    LaO molecule 178
    NH molecule 179
    nitrous oxide/acetylene flame 167
    NO molecule 180
    OH molecule 185, 211
    PO molecule 190
    SH molecule 197
    SiO molecule 198
    SnO molecule 205
Absorption coefficient 17

Absorption line profile 41
Accuracy 235
Active wavelength stabilization 149
AgH molecular absorption 155
Air/acetylene flame
    absorbance spectrum 187
    overview spectrum 154
    transmittance spectrum 149, 150
AlCl molecular absorption 158
AlF molecular absorption 160
AlH molecular absorption 162
Alkali halides 30
Alternate lines 61
Aluminum (Al) 94
    analysis 265
    determination in seawater 251
Ammonium nitrate
    modifier 253
Analyte addition technique 74
Analyte loss 235
Analytical line 59
    less sensitive 61, 237
Angular dispersion 19, 21
    of an echelle grating 36
Animal food
    analysis 215
Antimony (Sb) 97
Arc lamp 31
Array echelle spectrograph (ARES) **46**, 52, 72, 149, 153

Arsenic (As) 98
    determination in aluminum 265
    determination in urine 237
AsO molecular absorption 164
Atom mass 7
Atom release
    kinetic effects 236
Atomic (optical) emission spectroscopy
        (OES) 1
Atomic absorption
    general principle 17
    instrument effects 18
Atomic absorption spectrometry (AAS)
        3, 6, 270
Atomization curve 225
Atomization temperature 225
    optimization 232, 257
    optimum 91

Background absorption 59, 77
    by molecular dissociation 30
    excessive 231, 256
Background correction (BC) 23, 34, 40,
        58, **79**, 247
    deuterium (D_2) **77**, 213, 237
    erroneous 147
    high-current pulsing 78, 237
    least-squares 83, 86, 88, 99, 129,
        189, 211, **213**, 215, 217, 229,
        237, 240, 242, 253, 267
    measurement errors 78
    simultaneous 56, 225
    Smith-Hieftje 78, 237
    Zeeman-effect **77**, 237, 251, 253,
        265
Background correction pixel (BCP) 81,
        266
    correct choice 82

Background measurement 79
Background offset correction (BOC) 79
Barium (Ba) 98
Barycenter
    of intensity 151
Baseline adaptation
    automatic 56
Beer's law 17, 20
Beryllium (Be) 99
Binning
    vertical 52
Biological material
    analysis 245
Biological significance
    of the elements 91
Bismuth (Bi) 99
    determination in copper 217
    influence of the OH molecular
        absorption 213
Blaze angle 35, 36
Boron (B) 101
Bovine liver
    analysis 248
Broadband emission 58

Cadmium (Cd) 102
    absorbance noise 72
    determination in biological
        materials 247
    determination in coal 259, 260
    determination in copper 217
    determination in drinking water 213
    determination in urine 237, 240
Calcium (Ca) 103
Calibration curve 41, 63
Calibration function 62
Calibration range 63
Calibration sample 75

Cast iron
    analysis 223
Center pixel registration 65
Cesium (Cs) 103
Characteristic concentration 62, **69**, 91
Characteristic mass 62, 91
Charge-coupled device (CCD) 51, 54
Chemical modifier 257
Chemical vapor generation 271
Chromium (Cr) 104
    determination in drinking water 213
CN molecular absorption 104, 166
Coal
    analysis 230, 256
Coal fly ash
    analysis 256
Cobalt (Co) 106
    determination in biological
        materials 248
Cold vapor AAS 121
Collimator 11
Collision broadening 7–9
Collision cross-section 8, 13
Collision effects 7
Collisional line width 8
Conductance
    geometrical 21, **31**, 34, 39
Contamination 94, 130, 144, **235**
Continuous background absorption 81
Continuum source (CS) 1, 31
Convolution
    of Gauss and Lorentz profiles 8
    of Gauss and Voigt profiles 14
    of rectangular and sinc^2 profiles 11, 12
Copper (Cu) 108
    doublet 11, 12
    high-purity 216

Crude oil
    analysis 260
CS molecular absorption 169
CuH molecular absorption 173

Damping constant 9
Detector 50
    characteristics 23
    four-quadrant 34
    influence on absorption signal 23
    multi-pixel 18
    quantum efficiency 24
    read-out noise 24, 51
    read-out time 37, 269
    solid-state 34
Deuterium (D_2) background correction 77, 213, 237
Deuterium (D_2) lamp 33, 153
Diatomic molecules **25**, 46, 147
    for the determination of the elements 271
Diffraction of radiation 11
    angle 35
    efficiency 34
Diffraction order 36
    separation 34, 37
Dilution error 74
Direct line overlap 85, 89
    between aluminum and arsenic 265
    between copper and zinc 145, 216
    between iron and zinc 145, 215
Direct solid sample analysis
        see Solid sample analysis
Dispersion
    angular 19, 21, **36**
    linear 11, 36
Dispersive slit illumination (DSI) 48
Dissociation continua 30

Index

Dogfish liver
 analysis 248
Dogfish muscle
 analysis 248
Doppler broadening 5–7
Doppler effect 6
Doppler line width 6
Doppler profile 12, 13
Double echelle monochromator (DEMON) 37, 52
Double-beam spectrometer 58
Drinking water
 analysis 213
Dynamic range
 of a detector 51

Echelle grating 35
 auto-collimation mode 35, 38
Echelle monochromator 37, 39
Echelle spectrograph 37
Echelle spectrometer 4, 34
Electron excitation spectra **25**, 77, 78
 of diatomic molecules 147, 211
Electronic transitions 24–26
Elements
 determination via diatomic molecules 126, 136, **219**, 223, 271
Emission interference 57
Emission line
 width 63
Energy
 thermal **26**, 27, 29
Entrance slit 11
Environmental relevance
 of the elements 91
Europium (Eu) 109
Excited state 5, 25

Fine structure 5

Fine-structured background 85, 88
Fine-structured spectra 147
Fish muscle
 analysis 247
Flame
 air/acetylene 149, **150**, 154, 187
 nitrous oxide/acetylene 154, 167
 statistical fluctuations 189
Flame noise 68
Flicker-noise 24, 213
Focal plane 11
Free spectral range (FSR) 36
Frequency shift 6
Full width at half maximum (FWHM) 6
 of the instrument function 149

GaCl molecular absorption 176
Gallium (Ga) 109
Gauss function 13
Gauss profile **6**, 10, 23
Geometrical conductance 34
Germanium (Ge) 110
Gold (Au) 111
Graphite furnace (GF) 30
Graphite furnace AAS
 for direct analysis of solid samples 236
 method development 224
Grating
 echelle 35
 free spectral range (FSR) 36
 master 35
 number of grooves 36
 ruled width 35, 36
Grating equation 36
Green chemistry 236
Gull egg powder
 analysis 247

Halogen lamp 31
Halogens
    determination 271
Harnly, J.M. 4, 271
Heisenberg's uncertainty principle 5
Hieftje, G.M. 4
High-current pulsing background
        correction 78, 237
High-purity copper
    analysis 216
Hollow cathode lamp (HCL) **2**, 31,
        64–68
    spectral radiance 33
Hot-spot
    jitter 34
    mode of operation 31
Human hair
    analysis 248
Hydride-forming elements
    mutual interferences 271
Hyperfine splitting 63
Hyperfine structure 5

Illumination time
    influence on absorbance noise 70
Imprecision 236
Incidence
    angle 35
Indium (In) 111
Inductively-coupled plasma mass
        spectrometry (ICP-MS) 67, 271
Inductively-coupled plasma optical
        emission spectrometry (ICP OES)
        67, 73
Instrument
    optimal bandwidth 23
    resolving power 23
    transmittance 19

Instrument effects 18
Instrument profile 11–13, 269
    rectangular **12**, 19, 22
Instrumental bandwidth
    for CS AAS 23
Integrated absorbance 22, **84**, 226, 236
Intensity fluctuations 82
Interferences
            see Spectral interference
Internal standard 73
Intrinsic mass 62
Ionization limit 30, 91
Iridium (Ir) 112
    permanent modifier 247, 259
Iridium (Ir) and tungsten (W)
    mixed permanent modifier 260
Iron (Fe) 112
    analysis 215

L'vov, B.V. 64, 74
Lamp
    arc 32
    deuterium (D_2) 33, 153
    halogen 31
    hollow cathode (HCL) **2**, 31, 33,
        64–68
    neon 43–45
    xenon 3, 31–34, 54
Lamp intensity
    fluctuations 59
Lanthanum (La) 114
LaO molecular absorption 178
Lead (Pb) 115
    absorbance noise 70
    analytical line 59, 253
    determination in biological
        materials 245
    determination in coal 257

Index

determination in drinking water 213
determination in sediments 253
determination in soil 253
influence of the OH molecular
    absorption 212
Least-squares background correction
    see Background correction (BC)
Lichen
    analysis 247
Lifetime
    of the lamp 34
    of undisturbed excited states 5, 6
Light source
    modulated 57
Limit of detection (LOD) 18, 34, **68**
    shot-noise dominated 21
Line broadening 5
Line core 9, 62
Line overlap
    due to Zeeman-splitting of atomic
        lines 253
Line profile 5
    Doppler 12, 13
    Gauss **6**, 10, 23
    Lorentz **6**, 10, 13–16, 23
    Voigt 8–10, 14, 21
Line reversal 1
Line shift 8, 60
Line source (LS) 2, **31**, 57
Line width
    collisional 8
    Doppler 6
    natural 6
    Voigt 9, **10**, 16
Line wings 9, 63
Linear dispersion 11, 41
Lithium (Li) 116
Littrow-mounting 38

Lobster hepatopancreas
    analysis 248
Lorentz broadening 5, 7
Lorentz profile **6**, 10, 13–16, 23
Lorentz shape 6
Lorentz shift 8

Magnesium (Mg) 117
Magnetic field
    influence on molecular absorption
        237
Main components
    determination 60
Manganese (Mn) 117
Massmann, H. 77
Matrix constituents
    influence on molecular absorption
        211
Matrix elements
    in the sample 92
Matrix matching 74, 211
Maxwell velocity distribution 6
Measurement error 211
    due to background correction 78
Measurement pixel 68
Measurement signal 23
Measurement uncertainty 24
Measuring time 21
Mechanical chopper 58
Mercury (Hg) 120
Method development
    for graphite furnace analysis 224
Modulation principle 57
Molar mass 7, **8**, 16
Molecular absorption spectra 24–30, 77, 92
Molecular background
    in flame AAS 211

Molecule
    dissociation spectrum 30, 77
    electronic transition 25
    orbital 25
Molybdenum (Mo) 121
Motion
    of free atoms 6
Multi-element AAS
    sequential 72
    simultaneous 52, **72**, 269
Mussel tissue
    analysis 247

Natural line width 6
Nd:YAG laser 42
Negative absorbance 189, 211
Neon glow discharge lamp
    for wavelength calibration 43–45
Neon spectrum 45
NH molecular absorption 179
Nickel (Ni) 122
    determination in crude oil 260
    secondary line 122
    volatile compounds in crude oil 263
Nickel porphyrins
    determination in crude oil 264
Nitrous oxide / acetylene flame
    absorbance spectrum 167
    overview spectra 154
NO molecular absorption 99, 129, 137, **180**, 217, 242
Noise 68
    absorbance 19, 68–72
    flicker 24, 213
    read-out 24, 51
    shot 19, 24
Non-metal
    determination 271

Non-resonance line 61

Occurrence
    of the elements 91
OH molecular absorption 95, 99, 109, 117, **185**, 211
Optical density 18
Order separation
    external 37
    in an echelle spectrometer 37
    internal 37, 47
Oscillator strength 18
Over-correction 78, **83**, 250
Overview spectra 46, 148
    of individual molecules 153–210
Oyster tissue
    analysis 248

Palladium (Pd) 124
    chemical modifier **227**, 232, 257
    determination in urine 237
Peak volume registration 64, 85
Petroleum
    analysis 260
Pharmaceutical products
    analysis 215
Phosphoric acid
    overview spectrum 191
Phosphorus (P) 125
    determination in cast iron 223
    determination in pine needles 223
    determination in super phosphate fertilizer 223
    determination via PO molecular absorption bands 126, 219
Photo-multiplier tube (PMT) 51
Photodiode array (PDA) 3, 51
Photoelectron 24
Photon absorption 25

291

Photon energy 5
Photon flux 20
Photoplate 51
Piezo-electrical actuator 34
Pig kidney
    analysis 247
Pine needles
    analysis 223
Pixel errors 82
Pixel size 270
Pixel width 18
Platform atomization 226
Platinum (Pt) 127
PO molecular absorption 26, 27, 79, 98, 99, 102, 108, 115, 122, 129, **190**, 240, 248
    effect of the magnetic field 237
Polyatomic molecule 29, 147
Potassium (K) 128
    determination in animal food 215
    determination in pharmaceutical products 215
Precision 68, 235
Profile
    instrument 11–13, **19**, 22, 269
    line, see Line profile
    $sinc^2$ 11, 12
Pyrolysis curve 225
Pyrolysis temperature 225
    optimization **231**, 256, 263
    optimum 91

Quantum efficiency
    of the detector 51
Quartz-halogen lamp 31

Radiance 19
    of the radiation source 68
    spectral 31–33

Radiant power 17, **19**, 24
    absorbed 18
    incident 18
    minimum detectable decrease 20
    transmitted 17
Radiation scattering 77
Radiation source 31
    intensity 68
    stability 70
Rapidly changing background signal 79
Rare-earth elements 91
Rayleigh criterion 36
Read-out noise
    of a detector 24, 51
Read-out time
    of a detector 37, 269
Reduced palladium
    as modifier for nickel and vanadium 264
Reference element technique 73
Reference method 271
Reference pixel **68**, 70, 266
Reference spectrum 34, 56, **147**
    of interfering molecules 85, 229
Relative aperture 11
Reliability factor 20
Resolution
    of the spectrograph 48
    of the spectrometer 36
    spectral **34**, 148, 240, 246
Resolving power 36, 41
    diffraction-limited 36
    instrumental 34, 36, **38**
    of the spectrometer 34
Resonance transition 5
Rhodium (Rh) 128
Rotational fine structure 77
Rotational spectra 28

Rubidium (Rb) 128
Ruthenium (Ru) 129
    permanent modifier **245**, 253, 257
Rydberg series
    of aluminum 91

S_2 molecular absorption 138, 229
Sample preparation 235
Saturation capacity 270
Schumann-Runge bands 130
Seawater
    analysis 251
Secondary line 61, 73, **91**
Sediment
    analysis 227, 253
Sediment reference material
    molecular absorbance spectrum 200
Selectivity
    of atomic absorption measurements 57
Selenium (Se) 129
    absorbance noise 72
    determination in urine 237, 242
Self-absorption 60
Sensitivity 18, **62**, 64, 91, 235
Sequential multi-element AAS 72
Sequential spectrometer 37
SH molecular absorption 197
Shift
    of spectral lines 8
Shot-noise 19, 24
    dominated limit of detection 21
Side pixel registration 67
Signal distortion
    due to rapidly changing background signal 225
Signal processing
    in HR-CS AAS 79

Signal-to-noise ratio (SNR) 34, 59, **68**
    dependence on instrument profile 22, 23
Silicon (Si) 130
Silver (Ag) 133
Simultaneous background correction 56, 79, 225
Simultaneous determination
    of the elements 92
Simultaneous double-beam concept 58, 225
Simultaneous HR-CS AAS 52, 269
Simultaneous ICP OES 67, 73
Simultaneous multi-element AAS 3, 72
Simultaneous spectrometer 46
Simultaneous spectrum recording 148
$sinc^2$ profile 11, 12
Single-beam spectrometer 58
SiO molecular absorption 198
Slit
    entrance 11, 12, **18**, 39
    intermediate 39
Sloping background 83
Slurry sampling 253, 260
Smith-Hieftje background correction 78, 237
SnO molecular absorption 205
Sodium (Na) 133
    determination in animal food 215
    determination in pharmaceutical products 215
Soils
    analysis 253
Solid sample analysis 61, 121, 230, **235**, 245, 247, 248, 253, 257, 259
Solid-state array detector **51**, 148, 271
Species
    of elements 91

Specificity
    of atomic absorption measurements 57
Spectral interference 34, 56, 61, 63, 92, **147**, 266
Spectral line broadening 5
Spectral line profile
    see Line profile
Spectral radiance 31–33
Spectral range 148
Spectral resolution **34**, 148, 240, 246
Spectrometer
    bandwidth 19
    general requirements 34
    sequential 37
    simultaneous 46
    wavelength adjustment 38
Spectrum illumination time 269
Spectrum stabilization
    active 151
Stability of solutions 91
Stabilization
    of the spot image 34
    of the wavelength **34**, 49, 85, 152
Stabilized temperature platform furnace (STPF) 84, **226**, 236
Standards
    for calibration 236
Steel
    analysis 215
Stray light **21**, 22, 63
Strontium (Sr) 135
Sulfur (S) 135
    determination in cast iron 223
    determination via CS molecular absorption bands 136, 223
Sulfuric acid
    absorbance spectrum 169

Super phosphate fertilizer
    analysis 223
SuperDEMON 41
    instrument profile 13, 14

Tellurium (Te) 137
Tetramethylammonium hydroxide (TMAH)
    solubilization 248
Thallium (Tl) 138
    determination in coal 230, 256
    determination in coal fly ash 257
    determination in complex matrices 227
    determination in marine sediment 229, 253
Thermal drift 34
Thermodynamic equilibrium 6, 8
Time resolution 246
    of atomic and molecular absorption 230
Time-integrated absorbance spectrum 84
Tin (Sn) 139
Titanium (Ti) 141
Toxicity
    of the elements 91
Trace elements 216
    determination in drinking water 213
Transport interference 73
Transversely heated graphite tube 226
Trueness
    of the results 236
Tungsten (W) 141
Tungsten (W) and iridium (Ir)
    mixed permanent modifier 260
Tungsten halogen lamp 154

Under-correction 78
Urine
    analysis 237

Vacuum-UV 98
Valence electron 30
Vanadium (V) 142
    determination in crude oil 260
    volatile compounds in crude oil 263
Vanadyl porphyrins
    determination in crude oil 264
Vertical binning 52
Vibrational spectrum 26
Virtual CS AAS spectrometer **18**, 19, 21
Voigt line width 9, **10**, 16
Voigt profile 8–10, 14, 21
Volatility of compounds 91

Walsh, A. 1, 57, 74
Wavelength calibration
    accuracy 43
    self-controlling 42
Wavelength drift 149
Wavelength increments
    recorded simultaneously 269
Wavelength integrated absorbance (WIA)
    63
Wavelength selected absorbance (WSA)
    63
Wavelength stabilization 85, 152
    active 34, 49
Working curve
    non-linear 59
Working range **62**, 73, 237, 261
    linear 63, 64

Xenon lamp 3
    diffuse mode 31, 32
    hot-spot mode 31, 32
    lifetime 34
    operation parameter 32
    short-arc 31, 54

Zeeman-effect background correction **77**, 237, 251, 253, 265
Zeeman-splitting 253
Zinc (Zn) 144
    determination in copper 217
    determination in drinking water 213
    determination in iron and steel 215